Plain-English Study Guide for the General Radiotelephone Operator License (GROL)

by Richard P. Clem

Illustrated by Yippy G. Clem

PO Box 14957

Minnneapolis, MN 55414

INTRODUCTION

This study guide will prepare you for the FCC General Radiotelephone Operator (GROL) license. This book covers the Element 3 exam. To get your GROL, you also need to pass the Element 1 exam. If you already have any of the following licenses, then you have already passed Element 1:

Any level of commercial radiotelegraph license

Marine Radio Operator Permit

GDMSS Radio Operator's License (standard or restricted)

Since you are preparing for the more difficult Element 3 exam, then you probably need to do only minimal study for the Element 1 exam, since it consists of mostly much simpler material. However, you should take some time to study for the "easy" Element 1 test, since it does cover some marine communication information with which you might be unfamiliar. If you are starting from scratch, then I recommend my companion book, *Plain-English Study Guide for the Marine Radio Operator Permit*. That book follows the same format as this book and will prepare you for the Element 1 exam.

On the other hand, most students studying this book will find the MROP exam to be extremely simple. After you complete this book, you can simply read through the MROP questions on the FCC website. You'll probably get almost all of them correct, but if you miss a few, you should be able to remember the correct answer.

The GROL exam covers a wide variety of topics, from very basic DC and AC theory, to very specialized concepts such as air and sea navigation systems, RADAR, pulse modulation, and many more. Chances are, if you are planning on taking the test, you already have a good understanding of many of these areas. On the other hand, very few people work on a daily basis with **all** of the technology covered on the test. A broadcast engineer rarely works with RADAR. An avionics technician rarely deals with most marine equipment. And a marine electronics technician rarely deals with aviation concepts such as VOR,

DME, or aviation RADAR transponders.

Therefore, as you read this book, you will probably be able to gloss over many of the chapters. And when you are reading a chapter that deals with your own particular specialty, you will probably realize that I am oversimplifying much of the material. To that charge, I plead guilty. This book does not attempt to give you a solid understanding of any particular technology. Instead, my goal is to give you enough information to pass the test. If you're an avionics technician, then you almost certainly know a lot more than I do about avionics. If you're a RADAR technician, you almost certainly know a lot more about RADAR than I do. This book is not designed to make you a better avionics technician or RADAR technician. The book is designed with one goal in mind: To help you pass the test required in your particular field. And since that test requires you to know material from other fields, this book will teach you enough about those fields to pass the test.

The first three chapters (and some of the other chapters) cover some very basic concepts about DC electricity. For most people reading this book, those chapters will be very elementary, and you can probably gloss over them as well. On the other hand, the test includes a lot of questions on these basic concepts, and you do need a solid understanding of them. Therefore, I encourage you to read those chapters, even though they are very elementary. At the very least, you want to make sure that you can answer all of the questions in those chapters.

The chapters on AC theory will be helpful to many. AC theory can involve some fairly complicated mathematics. Fortunately, however, all of the questions on the exam can be answered by using some shortcuts, and I've presented those shortcuts here. If you understand the math and can answer the questions the "right way", then that's the way you should do it. But you'll pass the test by following my shortcuts.

Some topics have only a few questions, and rather than present the material in depth, I simply tell you what you need to memorize to get these questions right. But even in those cases, I give a short explanation of why the answer is right. Or, at the very least, I will give you a reminder that will help you memorize the answer based upon what you already know.

Each chapter in this book is followed by the actual questions that will appear on the exam. The correct answer to each question is shown in bold print. There are a total of 600 possible questions, and this book contains all 600. Each question contains the question number from the official question pool. These numbers are included in case you want to research other explanations of the question, some of which can be found on the Internet.

The actual test will consist of 100 of these questions, and you will need to answer 75 of them correctly. There is no penalty for wrong answers, so if you don't know the answer, the best strategy is to guess. You can almost always eliminate one or more wrong answers, which will increase your odds of guessing the right answer. The questions included here are from the question pool approved on June 25, 2009. These questions are in effect as of the date of publication, 2013. If you are reading a used copy of this book, you should check the FCC website to make sure that this is the current question pool. You can find this information at the following URL:

http://wireless.fcc.gov/commoperators/index.htm?job=home

At that same website, you can find the Element 1 question pool, if you still need to take that exam. The actual exams are conducted by a private contractor called a COLEM, and information about COLEM's and their exam schedules can be found at that same link.

There might be a few occasions when you can't answer the questions, even after re-reading the section. But I guarantee you that those occasions will be few. But if they do arise, you might have a handful of questions where you should simply memorize the question and the answer. To pass the test, you merely have to supply the correct answer. If you can do that by memorization, you will still get the question right, and you will still pass the test. **If you do memorize, you must memorize the correct answer, not just the letter for the answer.** The questions on the test will be exactly as shown here. However, the possible answers will probably be in a different order.

Don't attempt to memorize all of the answers. For one thing, that will be overwhelming. I know that I couldn't possibly memorize 600 questions and answers. But a few concepts will be new to you, and you might not have a good understanding. For those handful of times when you just don't "get it", don't feel bad if you have to simply memorize the correct answer.

CONTENTS

INTRODUCTION i

1 BASIC DC ELECTRICITY AND 1
   ELECTRONICS

2 COMBINING COMPONENTS 23

3 ADVANCED OHM'S LAW PROBLEMS 27

4 MAGNETISM 30

5 GALVANIC CORROSION 33

6 PARTS PER MILLION 35

7 RADIO WAVES 38

8 SPREAD SPECTRUM 49

9 CAPACITORS AND INDUCTORS 51

10 DIODES 58

11 SEMICONDUCTORS 70

12 WAVEFORMS 82

13 AC ELECTRICITY, REACTANCE, 89
    IMPEDANCE, AND RESONANCE

14 TIME CONSTANTS 113

15 TRANSFORMERS 119

16 DECIBELS 123

17 OSCILLATORS 128

18 RECEIVERS 134

19 AMPLIFIERS 149

20 FILTERS 154

21   OPERATIONAL AMPLIFIERS   159

22   SINGLE SIDEBAND AND AM   163

23   FM   171

24   BANDWIDTH   178

25   ANTENNAS AND FEEDLINES   182

26   INTERFERENCE   191

27   BATTERIES AND POWER SUPPLIES   194

28   MOTORS AND GENERATORS   202

29   RMS AND PEP   206

30   SAFETY   213

31   TTL   220

32   MULTIVIBRATORS AND FLIP-FLOPS   230

33   MICROPROCESSORS AND MEMORY   235

34   AIR NAVIGATION AND   240
     COMMUNICATION

35   MARINE COMMUNICATIONS   251

36   RADAR   259

37   PULSE MODULATION   269

38   SATELLITES   272

39   REPAIR AND MAINTENANCE, TEST   282
     EQUIPMENT

40   INDUSTRY STANDARDS AND   290
     SPECIFICATIONS

41   MISCELLANEOUS   297

1: BASIC DC ELECTRICITY AND ELECTRONICS

As you probably know, everything in the universe is made of a basic building block called the atom. The atom, in turn, is made up of three infinitesimally small particles called protons, neutrons, and electrons. The protons and neutrons are located in the center of the atom, and typically never go anywhere. The electrons, on the other hand, are orbiting the center (like tiny planets around the sun) and can move about much more freely.

Each chemical element has a certain number of protons, a certain number of neutrons, and a certain number of electrons. Since the electrons are free to move about, they sometimes leave the atom and go elsewhere. Or, an extra electron might enter the atom. Therefore, there are some atoms that have too many electrons, or two few electrons.

Typically, an atom that has too many electrons is eager to get rid of the extra atom, and it will do its best to get rid of the intruder. Given the opportunity, it will try to unload its extra electron on another unsuspecting atom nearby, as if it were a hot potato.

Similarly, if an atom has too few electrons, it wants to find a replacement wherever it can, and will snatch an electron away from another nearby atom, given the opportunity.

Picture in your mind a chain of atoms, sitting side by side. Let's say that the atom on the far left has an extra electron that it doesn't want. And let's say that the atom on the right is missing one of its electrons.

Realizing that it has an extra electron, the atom on the left waits until the atom to the right isn't looking, and then hands the "hot potato" electron to its neighbor. This atom doesn't want a spare electron any more than the original atom did, so it does its best to get rid of it. The original atom, having started this whole process in the first place, is obviously aware of what is going on, so the second atom probably won't be too successful in giving it back.

So what the second atom does is wait until an opportune moment, and then give the electron to the atom sitting to its right.

This whole process is repeated until the extra electron has been passed all the way down the line. But the last atom, the one at the right, was missing an electron. So this atom is quite happy to get the replacement, and all is well again in the atomic world.

This whole process is what is called an electric current. There was an electric current flowing from the atom on the left, all the way to the atom on the right.

There is a measurement of how strong an electric current is, and this measurement is the **Ampere**, or **Amp** for short. One Amp is defined as approximately 6000000000000000000 (that is 6, followed by 18 zeros) electrons per second passing a certain point. (You do not need to know that number for the test–it's just to give you an idea that it's a **lot** of electrons.) This number of electrons is also known as a **Coulomb**. In other words, one Amp is the same as one Coulomb per second.

It is sometimes useful to think of electricity as being similar to water. Electricity consists of electrons flowing through a wire. For many purposes, this is the similar to water flowing through a pipe. Therefore, you can think of the electric current, in amps, as being similar to the number of gallons per second passing through a pipe: It's a measure of how fast the electricity is moving. The number of gallons is akin to the number of Coulombs.

Let's think about water pipe again for a moment, because the behavior of water flowing through a pipe is similar to the behavior of electricity flowing through a wire. If we want to know what size of pipe to order, we'll want to know how much water is flowing through it–how many gallons per minute. If there is more flow, we'll need a bigger pipe.

Similarly, if there is more electric current (more amps), we'll need a bigger wire to handle the current.

But when we put in the new pipe, we'll also need to know how thick the walls of the pipe need to be. If there is too much water pressure and the pipe is made out of flimsy material, then the pipe might burst. We need to know how much pressure there will be. Normally, this is measured in pounds per square inch. This figure tells us how much energy there is in each drop of water.

There is a similar concept in electricity, and that is **voltage**. Voltage, as you might have guessed, is measured in **volts**. The number of volts tells us how much energy there is in each of the electrons passing through the wire. Energy is measured in **Joules**, and one volt is defined as one Joule per Coulomb. Just because we have a high current does not mean that the voltage will be high. This is just like water. If you look at a river, there is a huge amount of water passing by, so the current is very high. But the pressure is very low. (But if you try to squeeze that same river into a small pipe, the pressure will get much higher, a concept we'll talk about in a few minutes.)

Another name for voltage is **electromotive force (EMF)**. The higher the voltage, the more energy there is in each electron–the more pressure there is. One term you need to know for the exam is "back EMF". This simply means a voltage that opposes the applied EMF.

Let's think about that water pipe again. The water pipe running down the street in front of your house is probably a foot or more in diameter, because it is supplying water to all of the houses on your street. A lot of water passes through it, so it needs to be large. The pipe running into your house has a smaller diameter, because less water needs to pass through it. The pipes running to the individual sinks in your house are even smaller, because they need to carry even less water.

If you tried to replace the pipe in front of your house with the small pipe, it would cause much resistance to the flow of water. There is a similar concept in electricity, called **resistance**. Resistance is the measurement of how much a particular substance blocks the flow of electricity.

Substances with a very low resistance (like the big pipe running in the street) are good at conducting electricity. We call these substances **conductors**. In general, metals are good conductors of electricity–they have very low resistance. To put it in scientific terms, a conductor is a substance that has many free electrons. Non-metallic substances, such as plastic, dry wood, rubber, etc., have an extremely high resistance (for most practical purposes, they have infinite resistance). These substances are called **insulators**. One question on the exam will ask you which substance is the best insulator at UHF frequencies. The correct answer is mica.

Gold and silver are among the best electrical conductors. Copper and aluminum follow close behind, and they are the most common substances used for electric wires. Other metals are generally fairly good conductors. A piece of these materials would generally have a resistance of close to zero. A substance with a low resistance is said to have a high **conductivity**. One of the questions asks you to list various metals in the order of decreasing conductivity. The correct answer to that question is silver, copper, aluminum, iron, and lead. In other words, silver is the best conductor, and lead is the poorest conductor of those listed. The resistance of a particular substance is inversely proportional to its cross-sectional area. In other words, if you reduce the area by half, the resistance will double.

Some other substances can conduct electricity to some extent, but not as well as a true conductor. For example, carbon will conduct electricity, but it is like a very small pipe, because it doesn't conduct electricity very well. It has a high resistance, but not as high as plastic or rubber. Water can also conduct electricity, particularly if it is salt water. But it is not nearly as good a conductor as copper or other metals.

The amount of resistance of a piece of material is measured in **Ohms**. A perfect conductor will have a resistance of zero ohms. For example, a piece of copper wire will have a resistance very close to zero ohms. An insulator, such as plastic, will have a very high resistance. This will be many millions of ohms, and for practical purposes, we can just say that the resistance is an infinite number of ohms.

When we assemble an electronic device, sometimes we need to introduce resistance to the circuit. We use an electronic component called a **resistor**. When you buy a resistor, you specify the number of ohms that you need. Some resistors will be one ohm or less, and others will be many million ohms. They are generally made out of carbon.

One special kind of resistor is called a potentiometer. This is sometimes called a "variable resistor", but the word "potentiometer" is also used. A potentiometer is a resistor whose resistance can be changed, usually by turning the knob. If you turn the knob all the way one direction, the resistance is low. If you turn the knob all the way in the other direction, the resistance is high. A potentiometer is usually the component used as a volume control on a radio receiver. With the volume turned all the way down, the resistance is high. With the volume turned all the way up, the resistance is low. But when you turn the knob, you are adjusting the resistance (the number of ohms) of the resistor/potentiometer.

Think once again of a pipe with water flowing through it. Let's assume that we have water flowing through a three-inch pipe. At some point, the pipe narrows, and flows into a one-inch pipe. The current flowing through the pipe is the same everywhere. If you have ten gallons per minute flowing through the big pipe, and the only place for it to go is into the smaller pipe, then you'll have ten gallons per minute flowing through the smaller pipe.

But the water in the one-inch pipe will have more pressure. You can see this from the following diagram:

The pump is pumping water through the system. There is ten gallons per minute flowing through the entire loop. But the one-inch pipe is under a lot more pressure than the three-inch pipe.

Electricity works the same way. Look at the diagram below. This is a battery, some wire, and a light bulb. As you are probably aware, the light bulb will light up when it's hooked up this way. The light bulb gives off light because it is using energy from the electricity. And because it is using energy, it resists the flow of electricity. Therefore, it is not a true conductor, and it is not a true insulator. Instead, it is a resistor. We can measure the resistance of the light bulb with an instrument called an **ohmmeter**. For example, this light bulb might have a resistance of 2 ohms:

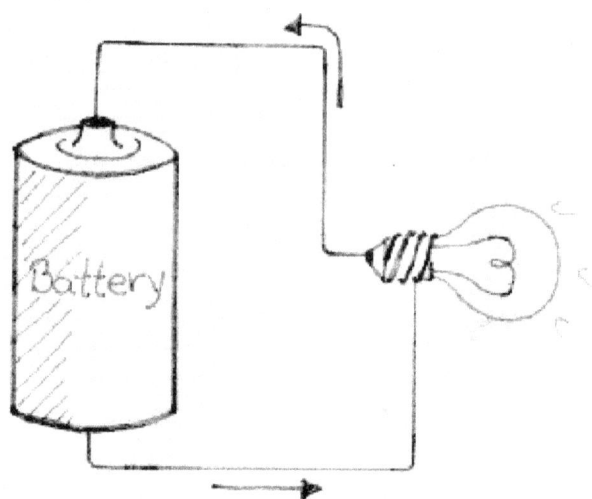

The light bulb in this diagram is like the one-inch pipe in the first diagram. Like the smaller section of pipe, it resists the flow of electricity. The battery in this diagram is like the pump in the first diagram. In the first diagram, the pump might be rated in terms of how much pressure it can generate–how many pounds per square inch. The battery is the same–it is rated in terms of how much electrical pressure (voltage) it puts out. Normal flashlight batteries (alkaline or carbon zinc) put out 1.5 volts. (We'll talk later about how batteries with larger voltages are put together).

As you are probably aware, normal alkaline or carbon-zinc batteries are not rechargeable. Some batteries are rechargeable. The most common ones for small electronic devices are **nickel-cadmium** (also known as **NiCd** or "**nicad**" batteries) or lithium-**ion** (also known as **LiIon** batteries). The voltage of a Nicad battery is 1.2 volts.

A normal car battery is known as a "lead-acid" battery. As you are aware, this type of battery can also be recharged, and is normally charged by your car's alternator.

Lead-acid batteries are commonly used to power equipment, but you must use caution around them, since they contain sulfuric acid. In addition, they can give off explosive hydrogen gas when being charged. They can overheat, or even explode, particularly if they are charged or discharged too quickly.

There is a relationship between voltage, current, and resistance (volts, amps, and ohms). This is a mathematical formula called **Ohm's Law**:

Volts equals amps time ohms.

Even if you don't understand this formula, you will be able to pass the test simply by memorizing the formula. If you are good with algebra, if you remember that first formula, you will be able to figure out the other two versions of the formula. But if you're not good with algebra, you can simply memorize the following two other versions of the formula:

Amps equals volts divided by ohms.

Ohms equals volts divided by amps.

When this formula is written, the letter "E" is often used to refer to voltage. ("E" stands for electromagnetic force, the other name for voltage.) "I" is often used to refer to current. (You can remember that "I" stands for the intensity of the current.) R, as you might expect, stands for resistance. Therefore, the three forms of Ohm's law can be written as follows:

$$E = I \times R$$

$$I = E / R$$

R = E / I

To understand these formulas, think again about the water flowing through the first diagram above. There's a certain amount of current (gallons per minute) flowing through the whole system. But when the water encounters the resistance of the small section of pipe, the pressure in that section goes up. We could assign a number for the resistance. Let's call it 10 "ohms" (of course, resistance of a water pipe isn't really measured in ohms, but it's close enough, that we're doing so for the purposes of this exercise. And then, we'll replace that section of pipe with a little bit larger one that is only five "ohms". So we have the following two setups:

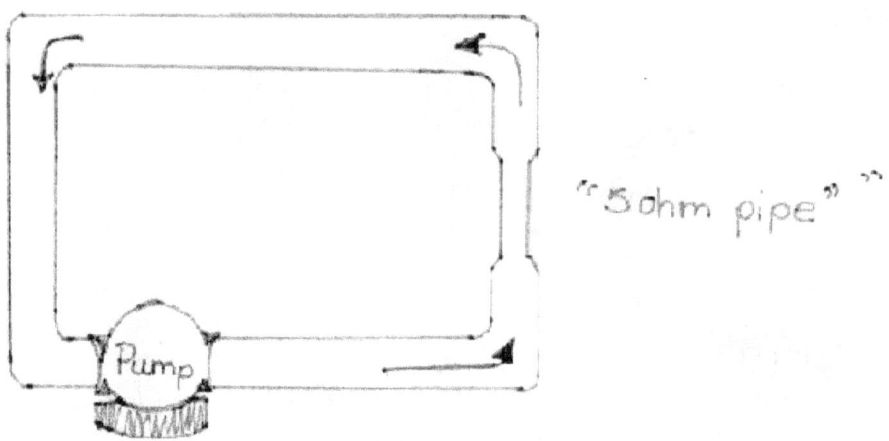

Under "Ohm's Law" (which doesn't really apply here, but again, it's close enough to demonstrate the point), the "voltage" (water pressure) equals current times ohms.

So in the drawing with the "10 ohm pipe", let's say that the current is 10 gallons per minute. The pressure (voltage) would be current times resistance, or 10 x 10 = 100. (At this point, I'm not saying 100 of what, because that would obviously depend on a lot of factors. But for the sake of simplicity, we'll say that the pressure is 100 pounds per square inch.

In the other diagram, we replace the "10 ohm" pipe with the larger "5 ohm" pipe. This time, the pressure (voltage) will be current times resistance, or 10 x 5 = 50.

In other words, when the resistance goes down, the voltage also goes down.

Now, obviously, in the case of water, this is a gross over-simplification, and things don't work out quite this simply. But you get the idea: As long as everything else stays the same, then the voltage goes up when the resistance goes up. This might seem backwards to you when it comes to electricity. But it makes more sense when you think about the water pipes. And it also makes more sense if you think of it this way: The larger resistor "uses up" more of the voltage than the smaller resistor. Or you need more voltage in the case of a larger resistor. We usually say that there is a "voltage drop" across the resistor.

Before we go on to draw the same diagram using electrical components, we should introduce the symbols for various electronic components. The picture of the light bulb and battery above is adequate, but if we want to draw more complicated electrical circuits, my artistic abilities are limited, and it would be very difficult for me to draw everything I need to draw. So we use standard symbols to represent everything. These symbols are called **schematic** symbols, and the layout of a circuit using these symbols is called a **schematic** diagram.

Here is the same diagram we drew earlier, using pictures of a battery and light bulb:

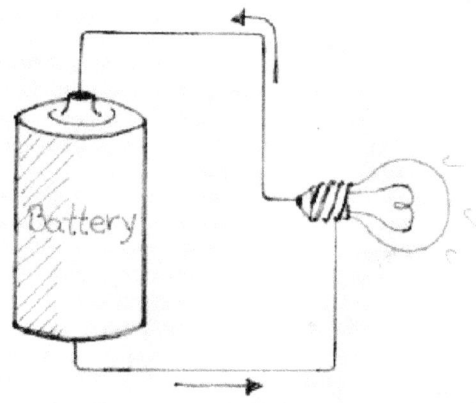

Here is the same diagram using these standard schematic symbols:

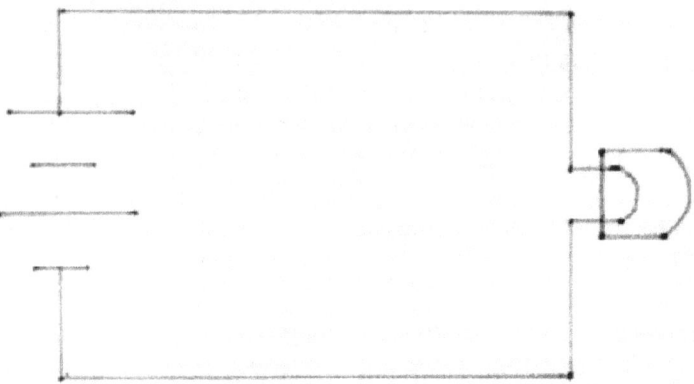

As you can see, we now know the symbols for a light bulb and for a battery:

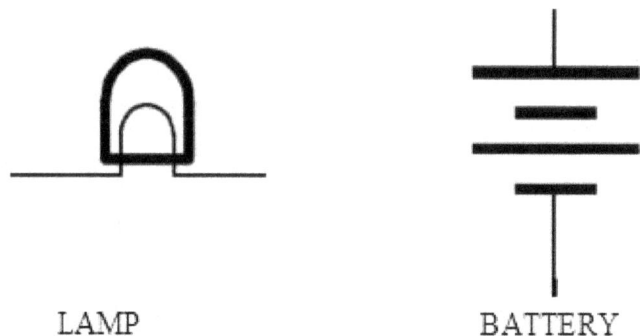

LAMP   BATTERY

And here is the symbol for a resistor. As you can see, it is just a zig-zag line:

RESISTOR

We previously talked about a potentiometer, also known as a variable resistor. Here is the schematic symbol for a variable resistor:

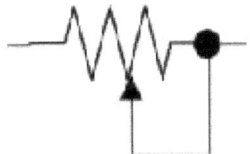

POTENTIOMETER (VARIABLE RESISTOR)

Now that we know some symbols, let's draw the circuit of a battery and a resistor:

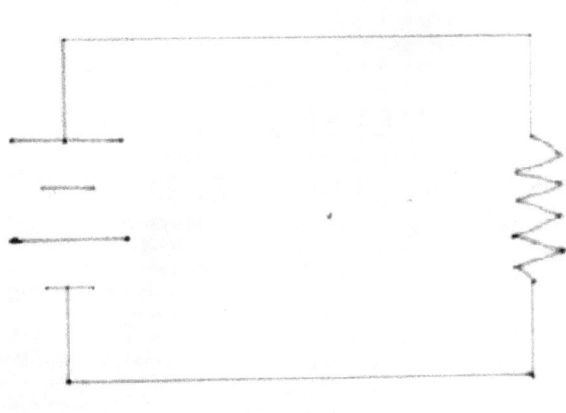

There are three things we can measure in this diagram. We can measure the voltage of the battery (the number of volts). We can measure the current flowing through the wire (the number of amps). And we can measure the resistance of the resistor (the number of ohms).

Using Ohm's law, as long as we know two of these numbers, we can always figure out the missing one.

For example, let's assume that the battery is 12 volts (such as a lead-acid car battery), and the resistor is 4 ohms. We can figure out the current in amps. Remember the formula:

Amps equals volts divided by ohms.

So here, there would be 3 amps, because 12 divided by 4 is 3.

For our next example, let's assume that the battery is 9 volts and the current is 4.5 amps. What is the resistance? Here, we remember the formula:

Ohms equals volts divided by amps.

So the correct answer is 2 ohms: 9 divided by 4.5.

Finally, we know that the current is 2 amps, and the resistance is 6 ohms. What is the voltage? Here, we remember the formula:

Volts equals amps times ohms.

So the battery is 12 volts: 2 times 6 equals 12.

There's one Ohm's law question that is cleverly disguised, because it asks for **conductance** rather than resistance. The unit of conductance is the **mho**, which you might have noticed is **ohm** spelled backwards. Conductance is merely one divided by resistance. So if the resistance is 2 ohms, the conductance is 0.5 mho. So for this problem, do the Ohm's law problem normally, and then take your answer and divide it into 1. (The mho is also called the **Siemen**, but both words are given on the exam, and mho makes it easier to remember that it's merely the inverse of ohm.)

A battery supplies electricity that is **direct current**. Direct current (DC) electricity is electricity that flows in one direction all the time. The electrons flow from the negative terminal of the battery, through the circuit, and back to the positive terminal of the battery. They never change directions. (Some people think of electricity as moving from the positive terminal to the negative terminal. It's moving so fast that it really doesn't matter which direction it's moving. The important thing is that DC is always moving the same direction.)

The other type of electricity is **alternating current**. Alternating current (AC) electricity changes direction on a regular basis. The outlets in your home are alternating current electricity. The electricity in your house changes direction 120 times per second. This is called 60 cycles per second, or 60 **Hertz** (abbreviated 60 Hz) electricity, because it goes through 60 complete cycles (it goes both directions) every second.

For the exam, you will need to be familiar with some of the metric prefixes. You will need to know **kilo-**, **Mega-**, **milli-**, and **micro-**, and **pico-**. "Micro" is represented by the Greek letter mu (μ). So a microsecond can be written as μs. You are probably already familiar with these, with the possible exception of the last one. A kilometer, for example, is a thousand meters. A kilogram is a thousand grams. "Megabucks" means a lot of money, in particular, one million dollars. A millimeter is 1/1000 of a meter.

"Pico" is a millionth of a millionth, or 1,000,000,000,000.

Here is the meaning of the ones that you will encounter on the test:

kilo	=	1000 of anything
Mega	=	1,000,000 of anything
milli	=	1/1000 of anything
micro	=	1/1,000,000 of anything
pico	=	1/1,000,000,000,000 of anything

For example, a kilowatt is 1000 watts. 4 megohms is 4,000,000 ohms. One millivolt is 1/1000 volt. One microfarad is 1/1,000,000 Farad. (Don't worry about the exact meaning of these terms at this point. For now, we're just concerned about understanding the meaning of these prefixes.)

Power is the rate at which electrical energy (or any other kind of energy) is being used. In general, when electricity passes through a resistance, it changes into heat. So power is also the rate at which the resistor is giving off heat (or perhaps light, or radio waves, or whatever the device is designed to do). Power is measured in **watts**. So a 100 watt light bulb uses electric energy (and gives of light and heat) twice as fast as a 50 watt light bulb.

Power is equal to voltage times current. Or, to put it another way:

Watts equals volts times amps.

If you know a little bit of algebra, you can figure out the other two versions of this equation:

Volts equals watts divided by amps.

Amps equals watts divided by volts.

Power is abbreviated as "P" in formulas. So this formula can be written as $P = EI$, $E = P/I$, and $I = P/E$. As with Ohm's law, any time you know two of these numbers (watts, amps, or volts), you can figure out the other one.

Horsepower is another unit by which power can be measured. For the exam, you should know that 746 watts is the same as one horsepower.

The device used to measure resistance is an **ohmmeter**. The device used to measure voltage is a **voltmeter**. The purpose of a series multiplier resistor in a voltmeter is to increase the voltage-indicating range.

The device used to measure current is an **ammeter**. The purpose of a shunt resistor in an ammeter is to increase the amperage-indicating range.

These three devices (voltmeter, ohmmeter, ammeter) are often included in one package, with a switch on the front to switch from one to the other. This combination is referred to as a **multitester**, or sometimes a "VOM". A digital version is called a **DMM** (digital multimeter). Some

DMM's have a "half digit". The purpose of this is to extend the accuracy at the lower part of the range.

If you are dealing with any type of analog meter, one important thing to consider is that the meter is most accurate toward the top of the scale. You should try to take readings only on the top 1/3 of the meter's scale, because this is the only part that is truly calibrated.

When measuring current, you need to put the ammeter in **series** with the circuit you are measuring. When you measure voltage, you keep the circuit hooked up, and measure the voltage **across** the battery or resistor ("in **parallel**" with the thing you are measuring). This measures the voltage of the battery, or the voltage drop across the resistor:

To measure resistance, the resistor must be removed from the circuit, and you connect one lead of the ohmmeter to each wire of the resistor:

When using a multitester, you need to make sure that it is switched to the proper type of reading. If you measure resistance and the thing you are measuring is hooked up to current, you will damage the meter. You will also damage the meter if it is set to measure current (amps) and you hook it directly across the battery.

3-1A2 What is the basic unit of electrical power?

    A. Ohm.

    B. Watt.

    C. Volt.

    D. Ampere.

3-1A3 What is the term used to express the amount of electrical energy stored in an electrostatic field?

    A. Joules.

B. Coulombs.

C. Watts.

D. Volts.

3-2A5 What is meant by the term "back EMF"?

A. A current equal to the applied EMF.

B. An opposing EMF equal to R times C (RC) percent of the applied EMF.

C. A voltage that opposes the applied EMF.

D. A current that opposes the applied EMF.

3-3A6 Which of these will be most useful for insulation at UHF frequencies?

A. Rubber.

B. Mica.

C. Wax impregnated paper.

D. Lead.

3-4A2 Good conductors with minimum resistance have what type of electrons?

A. Few free electrons.

B. No electrons.

C. Some free electrons

D. Many free electrons..

3-4A3 Which of the 4 groups of metals listed below are the best low-resistance conductors?

A. Gold, silver, and copper.

B. Stainless steel, bronze, and lead.

C. Iron, lead, and nickel.

D. Bronze, zinc, and manganese.

3-8A3 Halving the cross-sectional area of a conductor will:

A. Not affect the resistance.

B. Quarter the resistance.

C. Double the resistance.

D. Halve the resistance.

3-8A4 Which of the following groups is correct for listing common materials in order of descending conductivity?

A. Silver, copper, aluminum, iron, and lead.

B. Lead, iron, silver, aluminum, and copper.

C. Iron, silver, aluminum, copper, and silver.

D. Silver, aluminum, iron, lead, and copper.

3-9B1 What value of series resistor would be needed to obtain a full scale deflection on a 50 microamp DC meter with an applied voltage of 200 volts DC?

A. 4 megohms.

B. 2 megohms.

C. 400 kilohms.

D. 200 kilohms.

3-9B2 Which of the following Ohms Law formulas is incorrect?

A. $I = E / R$

B. $I = R / E$

C. $E = I \times R$

D. $R = E / I$

3-9B3 If a current of 2 amperes flows through a 50-ohm resistor, what is the voltage across the resistor?

A. 25 volts.

B. 52 volts.

C. 200 volts.

D. 100 volts.

3-9B4 If a 100-ohm resistor is connected across 200 volts, what is the current through the resistor?

A. 2 amperes.

B. 1 ampere.

C. 300 amperes.

D. 20,000 amperes.

3-9B5 If a current of 3 amperes flows through a resistor connected to 90 volts, what is the resistance?

A. 3 ohms.

B. 30 ohms.

C. 93 ohms.

D. 270 ohms.

3-13B3 746 watts, corresponding to the lifting of 550 pounds at the rate of one-foot-per-second, is the equivalent of how much horsepower?

A. One-quarter horsepower.

B. One-half horsepower.

C. Three-quarters horsepower.

D. One horsepower.

3-6I4 What is the total voltage when 12 Nickel-Cadmium batteries are connected in series?

A. 12 volts.

B. 12.6 volts.

C. 15 volts.

D. 72 volts.

3-6I5 The average fully-charged voltage of a lead-acid storage cell is:

A. 1 volt.

B. 1.2 volts.

C. 1.56 volts.

D. 2.06 volts.

3-6I6 A nickel-cadmium cell has an operating voltage of about:

A. 1.25 volts.

B.   1.4 volts.

C.   1.5 volts.

D.   2.1 volts.

3-64J1   What is the current flowing through a 52 ohm line with an input of 1,872 watts?

A.   0.06 amps.

B.   6 amps.

C.   28.7 amps.

D.   144 amps.

3-13B2   If a resistance to which a constant voltage is applied is halved, what power dissipation will result?

A.   Double.

B.   Halved.

C.   Quadruple.

D.   Remain the same.

3-18B6   What is the conductance (G) of a circuit if 6 amperes of current flows when 12 volts DC is applied?

A.   0.25 Siemens (mhos).

B.   0.50 Siemens (mhos).

C.   1.00 Siemens (mhos).

D.   1.25 Siemens (mhos).

3-74L1 What is a 1/2 digit on a DMM?

    A.   Smaller physical readout on the left side of the display.

    B.   Partial extended accuracy on lower part of the range.

    C.   Smaller physical readout on the right side.

    D.   Does not apply to DMMs.

3-74L2 A 50 microampere meter movement has an internal resistance of 2,000 ohms. What applied voltage is required to indicate half-scale deflection?

    A.   0.01 volts

    B.   0.10 volts

    C.   0.005 volts.

    D.   0.05 volts.

Hint: The question says that the meter should read half-scale. This means that for this simple Ohm's law problem, I = 25 microamps.

3-74L3 What is the purpose of a series multiplier resistor used with a voltmeter?

    A.   It is used to increase the voltage-indicating range of the voltmeter.

    B.   A multiplier resistor is not used with a voltmeter.

    C.   It is used to decrease the voltage-indicating range of the voltmeter.

    D.   It is used to increase the current-indicating range of an ammeter, not a voltmeter.

3-74L4 What is the purpose of a shunt resistor used with an ammeter?

A. A shunt resistor is not used with an ammeter.

B, It is used to decrease the ampere indicating range of the ammeter.

C. It is used to increase the ampere indicating range of the ammeter.

D. It is used to increase the voltage indicating range of the voltmeter, not the ammeter.

3-75L4 On an analog wattmeter, what part of the scale is most accurate and how much does that accuracy extend to the rest of the reading scale?

A. The accuracy is only at full scale, and that absolute number reading is carried through to the rest of the range. The upper 1/3 of the meter is the only truly calibrated part.

B. The accuracy is constant throughout the entire range of the meter.

C. The accuracy is only there at the upper 5% of the meter, and is not carried through at any other reading.

D. The accuracy cannot be determined at any reading.

2: COMBINING COMPONENTS

There are a few questions on the test about the value that results when you combine more than one component. In addition, there are a few questions in other chapters that also require you to combine components. For example, if you need a 200 ohm resistor but don't have one, you can combine two 100 ohm resistors in series:

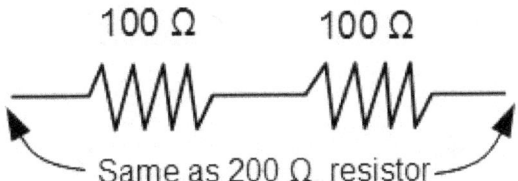

100 Ω   100 Ω

Same as 200 Ω resistor

Inductors work exactly the same way: They add up when connected in series.

Capacitors work differently. They add up when connected in **parallel**:

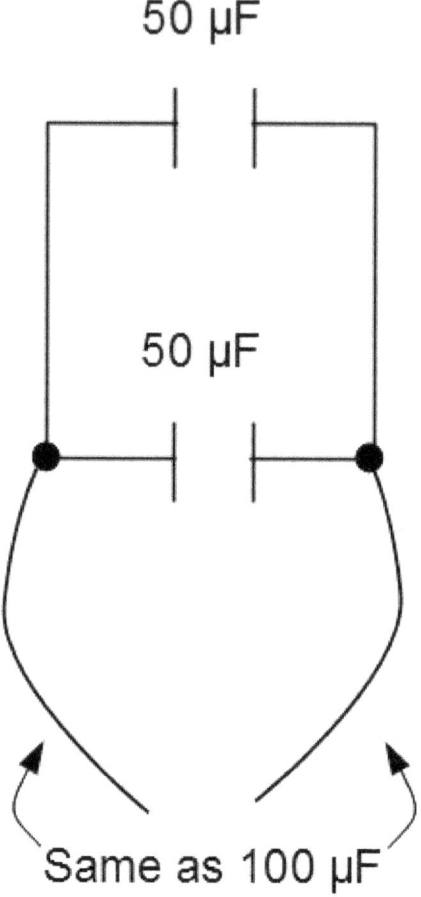

The math is slightly more complicated dealing with resistors in parallel. The first thing to remember is that the resistance always goes down: **The combined resistance of two or more resistors in parallel is always less than the smallest resistor.** If you remember this fact, you will be able to answer most of these questions. This makes sense, because the electricity has an additional path it can follow. Here are two resistors in parallel:

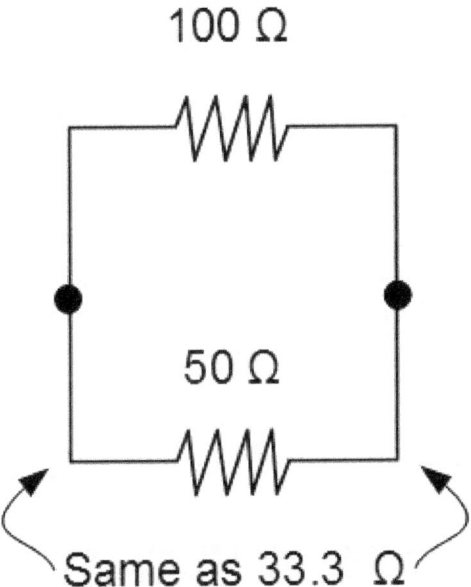

100 Ω

50 Ω

Same as 33.3 Ω

One resistor is 100 ohms, and the other one is 50 ohms. The combined resistance is less than the smallest resistor, and works out to 33.3 ohms. To calculate the combined resistance of **two** resistors in parallel, you use the "product over the sum" formula. You multiply the two values and get the product, in this case, 100 x 50 = 5000. The sum is the two numbers added together, 100 + 50 = 150. The final answer is the product divided by the sum: 5000 divided by 150, which equals about 33.3. So the answer is 33.3 Ohms.

This method only works with two resistors. If you have three or more resistors, you have to work on them two at a time. You can **not** multiply all three, and divide that by the sum or all three. If you have a question with more than one resistor in parallel, divide them into groups of two. Figure out the combined resistance of those two, and then combine it with the other one, using the same process.

Remember, inductors follow the same rules as resistors when it comes to combining them. So two inductors in parallel follow the "product over the sum" rule.

Capacitors in **series** also follow the "product over the sum" rule: If there are two capacitors in series, the combined capacitance will be less than any of the original capacitors.

3-4A1 What formula would calculate the total inductance of inductors in series?

    A. $L_T = L_1 / L_2$

    B. $L_T = L_1 + L_2$

    C. $L_T = 1 / L_1 + L_2$

    D. $L_T = 1 / L_1 \times L_2$

3-4A5 How would you calculate the total capacitance of three capacitors in parallel?

    A. $C_T = C_1 + C_2 / C_1 - C_2 + C_3$.

    B. $C_T = C_1 + C_2 + C_3$.

    C. $C_T = C_1 + C_2 / C_1 \times C_2 + C_3$.

    D. $C_T = 1 / C_1 + 1 / C_2 + 1 / C_3$.

3: ADVANCED OHM'S LAW PROBLEMS

Now that we know Ohm's law, and are able to combine components, we are able to do all of the Ohm's law problems on the test. When you encounter one of these problems, it is important to work methodically. Generally, you need to compute one value, and then use that information to compute the actual question given on the test. For example, you might be asked to calculate the voltage drop across a particular resistor. To do that, you'll first need to calculate the current passing through that resistor.

One of the problems includes a diode. Diodes will be discussed in another chapter. But as you're probably aware, a diode will conduct electricity only in one direction. In this problem, the diode is hooked up so that it does not conduct electricity. Therefore, it can be ignored:

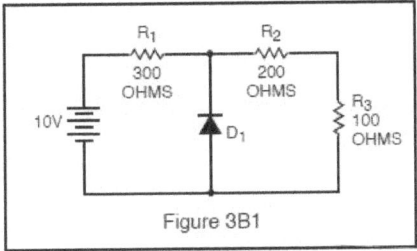

Figure 3B1

You are asked to calculate the voltage drop across R1. There are two ways to do this problem. As you can see, R1 is 300 ohms. Since we are ignoring the diode, all three diodes are in series. R2 + R3 = 300 ohms, so R1 is really in series with another 300 ohm resistor. Since both resistors are the same value, the voltage drop across each is equal, which means that half the voltage is being "used up" in each resistor. Therefore, the voltage drop across R1 is 5 volts.

If you didn't notice that the two resistors were equal, you could also do the problem by first calculating the current. The total value of the resistance is 600 ohms. Therefore, the current is $E/R = 10/600 = .01666$ amps. To calculate the voltage drop across R1, we use $E = IR = .01666$ x $300 = 5$ volts.

Some of the problems give the maximum power rating of a resistor, and ask what maximum voltage can be applied before the resistor burns out. For example, one question asks what is the maximum voltage that may be connected across a 20 watt, 2000 ohm resistor. From the power law, we know that $P = E^2/R$. In this case, $20 = E^2 / 2000$. With simple algebra (multiply both sides of the equation by 2000), we get $40,000 = E^2$. Then, we take the square root of both sides of the equation, and get $200 = E$. Therefore, the correct answer is 200 volts.

3-9B6 A relay coil has 500 ohms resistance, and operates on 125 mA. What value of resistance should be connected in series with it to operate from 110 V DC?

A. 150 ohms.

B. 220 ohms.

C. **380 ohms.**

D. 470 ohms.

3-10B2 What is the maximum DC or RMS voltage that may be connected across a 20 watt, 2000 ohm resistor?

A. 10 volts.   C. **200 volts.**

B. 100 volts.   D. 10,000 volts.

3-10B3 A 500-ohm, 2-watt resistor and a 1500-ohm, 1-watt resistor are connected in parallel. What is the maximum voltage that can be applied across the parallel circuit without exceeding wattage ratings?

A. 22.4 volts.

C. 38.7 volts.

B. 31.6 volts.

D. 875 volts.

3-10B4 In Figure 3B1, what is the voltage drop across R1?

A. 9 volts.

C. 5 volts.

B. 7 volts.

D. 3 volts.

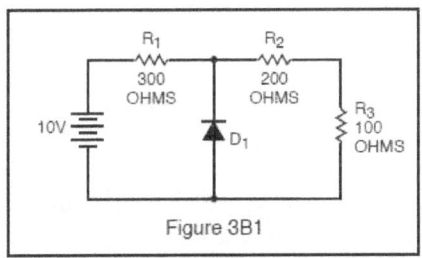

Figure 3B1

3-10B6 What is the maximum rated current-carrying capacity of a resistor marked "2000 ohms, 200 watts"?

A. 0.316 amps.

C. 10 amps.

B. 3.16 amps.

D. 100 amps.

4: MAGNETISM

You need some knowledge of magnetism, since it relates to other subjects, such as inductors, relays, motors, and generators. A magnetic field, as you probably know, can be produced by a current flowing through a conductor. The strength of the field is determined by the amount of the current.

You don't need to understand it in depth, but you do need to know what **Lenz's law** is. According to Lenz's law, induced currents produce expanding magnetic fields around conductors in a direction that opposes the original magnetic field.

Reluctance is the opposition to the creation of magnetic lines of force.

Finally, you need to know that **permeability** is defined as the ratio of magnetic flux density in a substance to the magnetizing force that produces it. If you remember simply that permeability is a ratio, then you will get this question right.

Speaking of magnets, there is one very simple question on the exam asking what a relay is. As you are probably aware, a relay consists of an electromagnet which is used to operate a switch. Since the switch can be far away from the operator, it's safe to say that this is a remotely controlled switching device, which is the correct answer to this question.

3-2A1 What determines the strength of the magnetic field around a conductor?

A. The resistance divided by the current.

B. The ratio of the current to the resistance.

C. The diameter of the conductor.

D. The amount of current.

3-2A2 What will produce a magnetic field?

A. A DC source not connected to a circuit.

B. The presence of a voltage across a capacitor.

C. A current flowing through a conductor.

D. The force that drives current through a resistor.

3-2A3 When induced currents produce expanding magnetic fields around conductors in a direction that opposes the original magnetic field, this is known as:

A. Lenz's law.

B. Gilbert's law.

C. Maxwell's law.

D. Norton's law.

3-2A4 The opposition to the creation of magnetic lines of force in a magnetic circuit is known as:

A. Eddy currents.

B. Hysteresis.

C. Permeability.

D. Reluctance.

3-2A6 Permeability is defined as:

A.   The magnetic field created by a conductor wound on a laminated core and carrying current.

B.   The ratio of magnetic flux density in a substance to the magnetizing force that produces it.

C.   Polarized molecular alignment in a ferromagnetic material while under the influence of a magnetizing force.

D.   None of these.

3-28C5 An electrical relay is a:

A.   Current limiting device.

B.   Device used for supplying 3 or more voltages to a circuit.

C.   Component used mainly with HF audio amplifiers.

D.   Remotely controlled switching device.

5: GALVANIC CORROSION

There are a few questions covering galvanic corrosion. **Galvanic corrosion** is corrosion resulting from electric current flowing between dissimilar metals. One question lists a number of objects on a ship, and asks which is the least affected by corrosion. These objects are made of aluminum, bronze, lead, and stainless steel. Of these, the stainless steel object (the propeller shaft) is the least affected by galvanic corrosion.

A sacrificial anode is often used to control galvanic corrosion. The most common material used is zinc.

3-3A1 What metal is usually employed as a sacrificial anode for corrosion control purposes?

    A. Platinum bushing.

    B. Lead bar.

    C. Zinc bar.

    D. Brass rod.

3-3A3 Which metal object may be least affected by galvanic corrosion when submerged in seawater?

    A. Aluminum outdrive.

    B. Bronze through-hull.

    C. Exposed lead keel.

    D. Stainless steel propeller shaft.

3-3A5 Corrosion resulting from electric current flow between dissimilar metals is called:

    A. Electrolysis.

    B. Stray current corrosion.

    C. Oxygen starvation corrosion.

    D. Galvanic corrosion.

6: PARTS PER MILLION

There are a number of questions about frequency accuracy that are quite easy as long as you understand the simple math. These questions ask how far off your frequency might be if you are within a certain number of parts per million.

The easiest way to approach these questions is to look at an example from the test: What is the most the actual transmit frequency could differ from a reading of 462,100,000 Hertz on a frequency counter with a time base accuracy of ± 0.1 ppm?

Here, "ppm" stands for "parts per million". If the accuracy is ± 0.1 ppm, this means it can be off by 0.1 millionth of the given value, either up or down.

The frequency given is 462,100,000, and we need to know how much 0.1 ppm of this value is. To do this, we simply multiply the original number by 0.1 and divide by one million:

462,100,000 x 0.1 / 1,000,000 = 46.21.

Therefore, the correct answer is 46.21 Hz. In other words, the actual frequency could be as low as 462,100,000 – 46.21 = 462,099,953.79, or it could be as high as 462,100,000 + 46.21 = 462,100,046.21.

These problems are very simple, and there is no reason to memorize the answers. If you do decide to memorize the answers, beware that the right answers for some questions are the same as the wrong answers for other questions.

3-11B1 What is the most the actual transmit frequency could differ from a reading of 462,100,000 Hertz on a frequency counter with a time base accuracy of ± 0.1 ppm?

A. 46.21 Hz.

B. 0.1 MHz.

C. 462.1 Hz.

D. 0.2 MHz.

3-11B4 What is the most the actual transmitter frequency could differ from a reading of 156,520,000 hertz on a frequency counter with a time base accuracy of ± 1.0 ppm?

A. 165.2 Hz.

B. 15.652 kHz.

C. 156.52 Hz.

D. 1.4652 MHz.

3-11B5 What is the most the actual transmitter frequency could differ from a reading of 156,520,000 Hertz on a frequency counter with a time base accuracy of +/- 10 ppm?

A. 146.52 Hz.

B. 1565.20 Hz.

C. 10 Hz.

D. 156.52 kHz.

3-11B6 What is the most the actual transmitter frequency could differ from a reading of 462,100,000 hertz on a frequency counter with a time base accuracy of ± 1.0 ppm?

A.   46.21 MHz.

B.   10 Hz.

C.   1.0 MHz.

D.   462.1 Hz.

7: RADIO WAVES

As you probably already know, radio communications takes place by using radio waves. The transmitter and antenna cause radio waves to be sent out in all directions. A radio wave is a type of "electromagnetic" wave. It has both an electrical and magnetic component.

Radio waves have a frequency. When you adjust the tuning dial on a normal AM or FM radio, you are adjusting the frequency that you are receiving. Different stations are using different frequencies. The frequency of a radio wave, just like an alternating current, is measured in cycles per second, or **Hertz (Hz)**. Since radio waves have a frequency in the millions of cycles per second, it is most common to refer to their frequencies in kiloHertz (kHz) (thousand cycles per second) or MegaHertz (MHz)(million cycles per second).

Radio signals also have a **wavelength**, which is measured in **meters**. Signals with a lower frequency have a long wavelength. Signals with a high frequency have a short wavelength. The frequency times the wavelength equals the speed of light, which is 300 million meters per second. (Note: You probably remember that the speed of light is 186,000 miles per **second**.

If you know the frequency, then you can calculate the wavelength, and vice versa. When you make this calculation, it is easiest to work in Megahertz, since you won't have so many zeros to worry about. If you're working in MHz, here are the formulas:

Wavelength = 300 divided by frequency.

Frequency = 300 divided by wavelength.

So a frequency of 50 MHz would have a wavelength of 6 meters: 300 / 50 = 6.

If you know the wavelength is approximately 40 meters and want to know the frequency, you would use the formula 300 divided by 40, which would mean the frequency is approximately 7.5 MHz.

The frequencies between 300 kHz and 3 MHz are called "MF" or "Medium Frequency". The frequencies between 3-30 MHz are called "HF" or "High Frequency". The frequencies between 30-300 MHz are called "VHF" or "Very High Frequency". The frequencies between 300-3000 MHz are called "UHF" or "Ultra High Frequency".

Some radio signals (usually HF signals, between 3-30 MHz) can travel long distances around the earth because they are reflected by the ionosphere.

The concept of harmonics is covered on the exam. This is a very simple concept. A **harmonic** of a frequency is simply a multiple of that frequency. For example, the second harmonic of 380 kHz is 760 kHz. It is called the "second harmonic" because we are multiplying the original frequency times two. The third harmonic of 380 kHz would be 1140 kHz. There is no such thing as the "first harmonic". The original frequency we started with is called the fundamental frequency.

There are some specific frequencies that are covered on the exam. You will need to know the following frequencies. For most of these questions, there is a wrong answer which includes the same numbers as the correct answer, but uses kHz instead of MHz, or *vice versa*. Therefore, you can't simply memorize the numbers. But if you learn the locations of these frequencies with respect to other frequencies you probably know, you'll get them right quite easily.

You will need to know that aircraft VOR (VHF Omni Range) and ILS (Instrument Landing System) operate on frequencies just above the FM broadcast band (just above 108 MHz.) VOR uses 108-117.95 MHz, and ILS uses 108.10-111.95. The ILS marker beacon, on the other hand, is just below the FM broadcast band, at 75 MHz.

Aircraft voice communications take place right above the VOR frequencies, at approximately 118-137 MHz.

Aircraft Distance Measuring Equipment (DME) operates in a frequency range of 962-1213 MHz. You won't have to remember the exact frequency, since the wrong answers are very different. If you remember that DME operates near 1000 MHz, you'll get this question right.

Automatic Direction Finding (ADF) receivers in both aircraft and ships use the frequency range 190-1750 kHz. ADF will be discussed in more detail in a later chapter. It uses LF and MF beacon transmitters at fixed locations to allow the receiver to obtain a bearing to that station. Standard AM broadcast stations can be used, so if you remember that the AM broadcast band (540-1700 kHz) is included in the correct answer, you will get this question correct on the exam.

Aircraft radar transponders transmit on 1090 MHz and receive on 1030 MHz. If you remember that the aircraft (high in the air) transmits on the higher frequency, you'll get this question right. Of course, you'll need to remember that this isn't taking place on the AM broadcast band, because the incorrect answers include 1030 and 1090 kHz.

Aircraft Emergency Locator Transmitters (ELT) operate on three frequencies: 121.5, 243, and 406 MHz. (If you remember that there are three such frequencies, then you'll get the question right, even if you don't remember the frequencies.)

An aircraft radio altimeter uses a frequency in the range of 4250-4350 MHz. It's not surprising that this device operates on microwave frequencies, and the correct answer for this question is the highest frequency of all of the possible answers given.

For the exam, you don't need to know the frequency range used by VHF marine radio, but you do need to know that 25 kHz channel spacing is used, and that the maximum allowable frequency deviation is +/- 5 kHz. You also need to know that VHF channel 16 is used for distress and calling, and that VHF channel 70 is used for Digital Selective Calling. To activate the DSC emergency signaling function on the VHF radio, a registered 9-digit MMSI (maritime mobile service identity) must be entered into the radio.

You need to know that many VHF marine radios have a "USA-INT" control. This control is used because some countries have different allocations, and the control changes **some** duplex channels to simplex. If you are confused by the wording of this question, just remember that it does not apply to all such channels, so select the only answer containing the word "some".

A search and rescue transponder (SART) is a device used aboard a lifeboat. If it detects a RADAR signal from another vessel, it transmits a signal that is visible on the RADAR screen. It appears as a series of 12 equally spaced dots. These devices operate at 9 GHz.

NAVTEX (Navigational telex) is a broadcast of navigational and weather information. The primary frequency for NAVTEX is 518 kHz, just below the AM broadcast band.

Satellites will be discussed in another chapter. For the exam, you should know that the Iridium satellite system operates on a frequency range of 1616-1626 MHz. (It might help you to remember that of the four possible answers, the correct answer is the smallest band, covering only 10 MHz.)

The Global Positioning Service (GPS) satellites operate on the frequencies of 1227.6 and 1575.4 MHz. If you are memorizing this answer, you can simply remember that these numbers end in .4 and .6.

3-11B2 The second harmonic of a 380 kHz frequency is:

A. 2 MHz.

B. 760 kHz.

C. 190 kHz.

D. 144.4 GHz.

3-11B3 What is the second harmonic of SSB frequency 4146 kHz?

A. **8292 kHz.**

B. 4.146 MHz.

C. 2073 kHz.

D. 12438 kHz.

3-68K1 What is the frequency range of the Distance Measuring Equipment (DME) used to indicate an aircraft's slant range distance to a selected ground-based navigation station?

A. 108.00 MHz to 117.95 MHz.

B. 108.10 MHz to 111.95 MHz.

C. **962 MHz to 1213 MHz.**

D. 329.15 MHz to 335.00 MHz.

3-69K3 What is the frequency range of the ground-based Very-high-frequency Omnidirectional Range (VOR) stations used for aircraft navigation?

A. 108.00 kHz to 117.95 kHz.

B. 329.15 MHz to 335.00 MHz.

C. 329.15 kHz to 335.00 kHz.

D. **108.00 MHz to 117.95 MHz.**

3-70K1 What is the frequency range of the localizer beam system used by aircraft to find the centerline of a runway during an Instrument Landing System (ILS) approach to an airport?

A. 108.10 kHz to 111.95 kHz.

B. 329.15 MHz to 335.00 MHz.

C.   329.15 kHz to 335.00 kHz.

D.   108.10 MHz to 111.95 MHz.

3-70K2 What is the frequency range of the marker beacon system used to indicate an aircraft's position during an Instrument Landing System (ILS) approach to an airport's runway?

A.   The outer, middle, and inner marker beacons' UHF frequencies are unique for each ILS equipped airport to provide unambiguous frequency-protected reception areas in the 329.15 to 335.00 MHz range.

B.   The outer marker beacon's carrier frequency is 400 MHz, the middle marker beacon's carrier frequency is 1300 MHz, and the inner marker beacon's carrier frequency is 3000 MHz.

C.   The outer, the middle, and the inner marker beacon's carrier frequencies are all 75 MHz but the marker beacons are 95% tone-modulated at 400 Hz (outer), 1300 Hz (middle), and 3000 Hz (inner).

D.   The outer, marker beacon's carrier frequency is 3000 kHz, the middle marker beacon's carrier frequency is 1300 kHz, and the inner marker beacon's carrier frequency is 400 kHz.

3-71K1 What is the frequency range of an aircraft's Automatic Direction Finding (ADF) equipment?

A.   190 kHz to 1750 kHz.

B.   190 MHz to 1750 MHz.

C.   108.10 MHz to 111.95 MHz.

D.   108.00 MHz to 117.95 MHz.

3-71K3 What are the transmit and receive frequencies of an aircraft's mode C transponder operating in the Air Traffic Control RADAR Beacon System (ATCRBS)?

A.   Transmit at 1090 MHz, and receive at 1030 MHz

43

B. Transmit at 1030 kHz, and receive at 1090 kHz

C. Transmit at 1090 kHz, and receive at 1030 kHz

D. Transmit at 1030 MHz, and receive at 1090 MHz

3-72K3 What is the frequency range of an aircraft's Very High Frequency (VHF) communications?

A. 118.000 MHz to 136.975 MHz (worldwide up to 151.975 MHz).

B. 108.00 MHz to 117.95 MHz.

C. 329.15 MHz to 335.00 MHz.

D. 2.000 MHz to 29.999 MHz.

3-72K4 Aircraft Emergency Locator Transmitters (ELT) operate on what frequencies?

A. 121.5 MHz.

B. 243 MHz.

C. 121.5 and 243 MHz.

D. 121.5, 243 and 406 MHz.

3-72K5 What is the frequency range of an aircraft's radio altimeter?

A. 962 MHz to 1213 MHz.

B. 329.15 MHz to 335.00 MHz.

C. 4250 MHz to 4350 MHz.

D. 108.00 MHz to 117.95 MHz.

3-81L6 What steps must be taken to activate the DSC emergency signaling function on a marine VHF?

    A.   Separate 12 volts to the switch.

    B.   Secondary DSC transmit antenna.

    C.   GPS position input.

    D.   Input of registered 9-digit MMSI.

3-85N1 What is the channel spacing used for VHF marine radio?

    A.   10 kHz.

    B.   12.5 kHz.

    C.   20 kHz.

    D.   25 kHz.

3-85N2 What VHF channel is assigned for distress and calling?

    A.   70

    B.   16

    C.   21A

    D.   68

3-85N3 What VHF Channel is used for Digital Selective Calling and acknowledgement?

    A.   16

    B.   21A

    C.   70

    D.   68

3-85N4 Maximum allowable frequency deviation for VHF marine radios is:

A. +/- 5 kHz.

B. +/- 15 kHz.

C. +/- 2.5 kHz.

D. +/- 25 kHz.

3-85N5 What is the reason for the USA-INT control or function?

A. It changes channels that are normally simplex channels into duplex channels.

B. It changes some channels that are normally duplex channels into simplex channels.

C. When the control is set to "INT" the range is increased.

D. None of the above.

3-87N1 What causes the SART to begin a transmission?

A. When activated manually, it begins radiating immediately.

B. After being activated the SART responds to RADAR interrogation.

C. It is either manually or water activated before radiating.

D. It begins radiating only when keyed by the operator.

3-87N2 How should the signal from a Search And Rescue RADAR Transponder appear on a RADAR display?

A. A series of dashes.

B. A series of spirals all originating from the range and bearing of the SART.

   C. A series of twenty dashes.

   D. A series of 12 equally spaced dots.

3-87N3 In which frequency band does a search and rescue transponder operate?

   A. 9 GHz

   B. 3 GHz

   C. S-band

   D. 406 MHz

3-88N5 Which of the following is the primary frequency that is used exclusively for NAVTEX broadcasts internationally?

   A. 2187.5 kHz.

   B. 4209.5 kHz.

   C. VHF channel 16.

   D. 518 kHz.

3-95P2 What frequency band is used by the Iridium system for telephone and messaging?

   A. 965 - 985 MHz.

   B. 1616 -1626 MHz.

   C. 1855 -1895 MHz.

   D. 2415 - 2435 MHz.

3-98P2 The GPS transmitted frequencies are:

A. 1626.5 MHz and 1644.5 MHz.

B. 1227.6 MHz and 1575.4 MHz.

C. 2245.4 and 2635.4 MHz.

D. 946.2 MHz and 1226.6 MHz.

8: SPREAD SPECTRUM

There are only a few questions on the exam regarding spread spectrum, and they are quite simple. Spread spectrum is defined as a wide bandwidth communication system in which the RF carrier varies according to a pre-determined sequence. It can use either frequency hopping of a single carrier over a pre-determined list of pseudo-random channels, or direct sequence, which is the use of multiple simultaneous carriers.

3-56G4 What term describes a wide-bandwidth communications system in which the RF carrier varies according to some pre-determined sequence?

    A. Spread-spectrum communication.

    B. AMTOR.

    C. SITOR.

    D. Time-domain frequency modulation.

3-56G6 What is the modulation type that can be a frequency hopping of one carrier or multiple simultaneous carriers?

    A. SSB.

    B. FM.

C. OFSK.

D. Spread spectrum.

3-82M1 What term describes a wide-bandwidth communications system in which the RF carrier frequency varies according to some predetermined sequence?

A. Amplitude compandored single sideband.

B. SITOR.

C. Time-domain frequency modulation.

D. Spread spectrum communication.

3-82M2 Name two types of spread spectrum systems used in most RF communications applications?

A. AM and FM.

B. QPSK or QAM.

C. Direct Sequence and Frequency Hopping.

D. Frequency Hopping and APSK.

3-82M3 What is the term used to describe a spread spectrum communications system where the center frequency of a conventional carrier is altered many times per second in accordance with a pseudo-random list of channels?

A. Frequency hopping.

B. Direct sequence.

C. Time-domain frequency modulation.

D. Frequency compandored spread spectrum.

9: CAPACITORS AND INDUCTORS

A capacitor is a device used to store electrical energy in an electrostatic field. The schematic diagram for a capacitor is shown below:

CAPACITOR

A simple capacitor consists of two metal plates separated by a non-conducting material, such as air. A capacitor has a property called capacitance, and this is measured in the unit Farad. The non-conducting material is called the **dialectric**, and has a property known as the **dialectric constant**. For the exam, you need to know that air has a relative dialectric constant of 1. The capacitance of a capacitor is determined by the size of the plates, the distance between the plates, and the dialectric constant of the material between the plates.

A **coupling capacitor** is often used between two stages of an amplifier or receiver. The coupling capacitor blocks direct current, but allows the AC signal to pass to the next stage. If a coupling capacitor in an amplifier fails, but does not short out (in other words, it "fails open"), then the result would be that no amplification would occur (because there would be nothing to amplify). The DC voltages in the circuit, however, would be normal.

Two of the capacitor questions on the test refer to the following diagram:

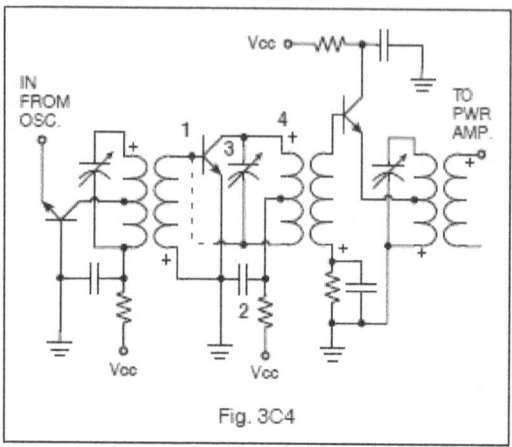

Fig. 3C4

Here is the material you need to know regarding this diagram. First of all, one could place a variable capacitor at the spot marked by the dashed lines in order to decrease parasitic oscillations.

The capacitor marked 2 is there to provide a signal ground. It provides a path to ground for the AC signal, but does not allow DC to pass.

Two questions refer to the following diagram of an audio amplifier:

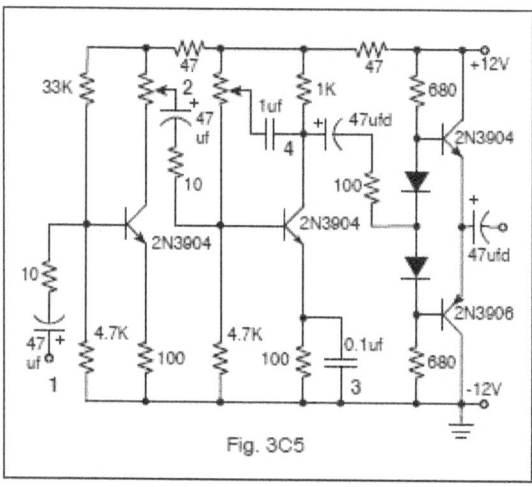

Fig. 3C5

The capacitor marked 3 is **a bypass capacitor**. This means simply that it removes alternating current by providing a low impedance path to ground. You will note that the 1 μF capacitor is connected to a potentiometer. This is the tone control. By adjusting the potentiometer, more of the signal is allowed to pass through the 1 μF capacitor. Since the capacitor will have a lower reactance for higher frequency signals, this allows more of the high frequency signals to pass through.

Because of its ability to pass RF signals but not DC current, a capacitor is often used to solve radio frequency interference problems. One question on the exam asks what might be done if an elevator's tachometer is experiencing interference from a nearby radio transmitter. The correct answer is to install a .01 μF capacitor across the motor/tachometer leads. When you think about it, this is a reasonable solution. The motor's DC will be unaffected, but the RF will be shorted out by the capacitor. A small capacitor such as the .01 μF in the correct answer would be more than adequate. One of the wrong answers calls for 200 μF, which would potentially cause other problems with the motor or tachometer.

An inductor is a device that stores energy in a magnetic field. Its inductance is measured in the unit **Henry**. As you probably know, a larger coil has a larger inductance, and a smaller coil has a smaller inductance.

More information about capacitors and inductors is contained in the chapter on reactance and impedance.

3-1A4 What device is used to store electrical energy in an electrostatic field?

    A. Battery.

    B. Transformer.

    C. Capacitor.

    D. Inductor.

3-3A2 What is the relative dielectric constant for air?

    A. 1

B. 2

C. 4

D. 0

3-4A4 What is the purpose of a bypass capacitor?

A. It increases the resonant frequency of the circuit.

B. It removes direct current from the circuit by shunting DC to ground.

C. It removes alternating current by providing a low impedance path to ground.

D. It forms part of an impedance transforming circuit.

3-4A6 How might you reduce the inductance of an antenna coil?

A. Add additional turns.

B. Add more core permeability.

C. Reduce the number of turns.

D. Compress the coil turns.

3-20C1 What factors determine the capacitance of a capacitor?

A. Voltage on the plates and distance between the plates.

B. Voltage on the plates and the dielectric constant of the material between the plates.

C. Amount of charge on the plates and the dielectric constant of the material between the plates.

D. Distance between the plates and the dielectric constant of the material between the plates.

The following two questions refer to this figure:

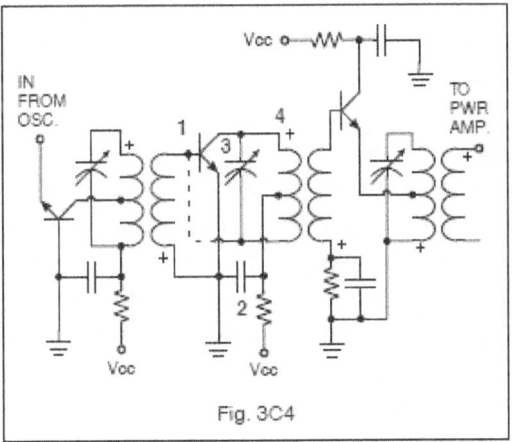

Fig. 3C4

3-20C2 In Figure 3C4, if a small variable capacitor were installed in place of the dashed line, it would?

A. Increase gain.

B. Increase parasitic oscillations.

C. Decrease parasitic oscillations.

D. Decrease crosstalk.

3-20C3 In Figure 3C4, which component (labeled 1 through 4) is used to provide a signal ground?

A. 1

B. 2

C. 3

D. 4

The following two questions refer to this figure:

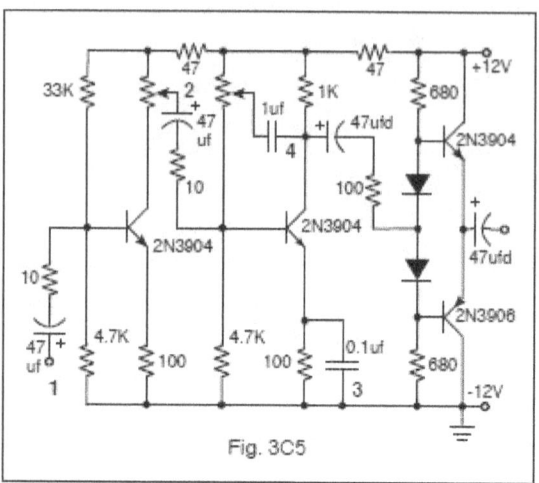

Fig. 3C5

3-20C4 In Figure 3C5, which capacitor (labeled 1 through 4) is being used as a bypass capacitor?

    A. 1

    B. 2

    C. 3

    D. 4

3-20C5 In Figure 3C5, the 1 µF capacitor is connected to a potentiometer that is used to:

    A. Increase gain.

    B. Neutralize amplifier.

    C. Couple.

    D. Adjust tone.

3-20C6 What is the purpose of a coupling capacitor?

A. It blocks direct current and passes alternating current.

B. It blocks alternating current and passes direct current.

C. It increases the resonant frequency of the circuit.

D. It decreases the resonant frequency of the circuit.

3-32D3 What will occur if an amplifier input signal coupling capacitor fails open?

A. No amplification will occur, with DC within the circuit measuring normal.

B. Improper biasing will occur within the amplifier stage.

C. Oscillation and thermal runaway may occur.

D. An AC hum will appear on the circuit output.

3-81L2 The tachometer of a building's elevator circuit experiences interference caused by the radio system nearby. What is a common potential "fix" for the problem?

A. Replace the tachometer of the elevator.

B. Add a .01 µF capacitor across the motor/tachometer leads.

C. Add a 200 µF capacity across the motor/tachometer leads.

D. Add an isolating resistor in series with the motor leads.

10: DIODES

A **diode** is a semiconductor device that allows current to flow in only one direction. The schematic symbol for a diode is shown below:If a diode is

 DIODE

forward biased, this means that it's hooked up so that the current can flow. Another way of saying this is that a diode has low impedance when it is forward biased. In other words, the arrow is connected to the positive voltage, and the line is connected to the negative voltage. Another way to look at it is that the current is flowing (from positive to negative) in the direction of the arrow.

One of the questions on the exam involves the following diagram, and asks you to tell which lamps will be lit:

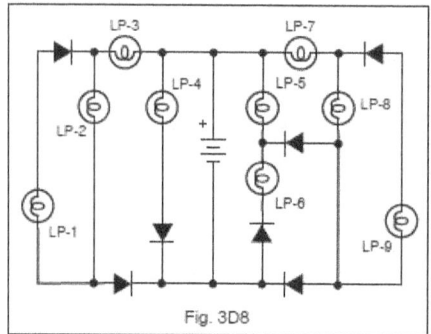

Fig. 3D8

To do this problem, you simply trace the path of electricity from the positive terminal of the battery to the negative terminal. We can quickly eliminate some wrong answers by seeing that lamp 4 is definitely on. The current comes from the positive terminal of the battery, to the lamp, through the diode (in the direction of the arrow) and back to the negative terminal of the battery. You simply repeat this process with the remaining lamps until you've eliminated all of the wrong answers. Remember, the bulbs can light even if more than one are wired in series, so the current to one lamp might pass through another lamp.

One question asks you to match up the output waveform from the following circuit:

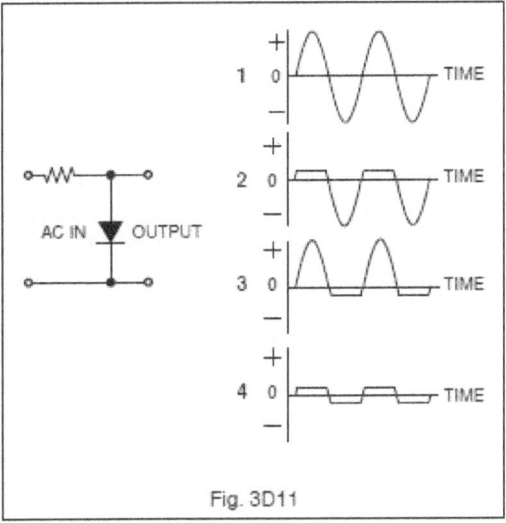

Fig. 3D11

The answer is waveform #2. If you stop and think, it is clear that this is the right answer. When the input is positive, then all of the current passes through the diode, and the output voltage drops to nearly zero. But when the input is negative, the diode has no effect, and the output passes through without change.

When selecting a diode, the two most common specifications are the maximum forward current and the PIV (peak inverse voltage). You should know for the exam that the main factor limiting the forward current is the junction temperature.

In a receiver, a diode is used as a detector. Detection is the process of rectification and filtering of the RF.

Structurally, a diode can be of one of two types. They are the junction diode and the point contact diode.

There are a number of special types of diodes covered by the test. The first of these is the **zener diode**.

A zener diode can function the same way as a normal diode, but it also allows current to flow in the reverse direction, as long as the voltage is above a voltage known as the zener voltage. Because the voltage at which this happens can be known very precisely, zener diodes are often used as reference voltage sources for voltage regulators or power supplies. Voltage ratings can vary from between 2.4 and 200 volts.

The zener diode is hooked up "backwards" in the circuit, so that the line (which looks somewhat like a "Z" rather than a straight line) is hooked to the positive voltage, and the arrow is hooked to the negative voltage. A zener diode will have a maximum and minum current rating, and a series resistor is normally used to keep the resistance in this range. One question asks you to compute the value of a series resistor for a zener diode with a current range of 5-95 mA. The question specifies that the zener diode has an internal impedance of 8 ohms. The supply voltage is 20 volts.

Since we want the current to be in about the middle of this range, we can pick a current at about the middle of the range. Let's call this current 50 mA, or .05 amps. This is now a simple Ohm's Law problem: We know current and voltage, and want to find resistance. From Ohm's Law:

$$R = E / I$$

$$R = 20 / .05 = 400 \text{ Ohms}$$

We know that the internal impedance is 8 ohms, so we need an additional 392 ohms. The closest answer on the test is 200 ohms. This is somewhat far away from what we calculated, but we have a wide range to work with. To make sure this is the correct answer, let's do the problem the other way, and make sure that the current falls in the correct range. Remember, assuming that 200 ohms is the right answer, the total resistance is 208 ohms, since the zener diode has internal resistance of 8 ohms.

I = E / R

I = 20 / 208 = .096 amps = 9.6 mA.

This is well within the range of 5-95 mA, so 200 ohms is the correct answer.

Finally, one question asks you to compute the voltage drop across R1 in the following diagram:

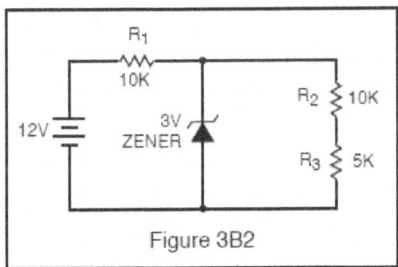

Figure 3B2

This problem is very simple. The problem doesn't state the current range of the zener diode, so we must assume that it is operating as a zener diode, which means that the voltage drop across R2 and R3 is exactly 3 volts. Since that voltage drop is 3 volts, the remaining voltage drop across R1 must be 9 volts, which is the correct answer.

Another special type of diode covered by the test is the **light-emitting diode (LED).** As the name implies, this diode gives off light when current is passing through it. To give off light, the current must be flowing. Therefore, we say that the LED is forward biased. The color of the LED depends upon the materials used in its construction.

The current used by most LED's is about 20 mA. Since an LED is essentially a short circuit when it's hooked up (current flows freely through it), it cannot be hooked directly to a battery. Instead, a series resistor is hooked up.

One component that contains an LED is called an **optoisolator** or **optocoupler.** This device is used to link two circuits together with no electrical connection. It consists of an LED and some type of photosensitive device. The LED is connected the output of one circuit, and the photosensitive device serves as the input of the other circuit.

Other special type of diode covered by the test are:

The PIN diode. They are commonly used as an RF switch.

The hot-carrier diode. They are commonly used as VHF and UHF mixers and detectors.

The tunnel diode. A tunnel diode is capable of both amplification and oscillation. It is capable of doing this because it has a region of negative resistance.

The varactor diode. A varactor diode has an internal capacitance that can be varied depending on the voltage. Therefore, they can be used for tuning circuits.

3-5A1 What are the two most commonly-used specifications for a junction diode?

A. Maximum forward current and capacitance.

B. Maximum reverse current and PIV (peak inverse voltage).

C. Maximum reverse current and capacitance.

D. Maximum forward current and PIV (peak inverse voltage).

3-5A2 What limits the maximum forward current in a junction diode?

A. The peak inverse voltage (PIV).

B. The junction temperature.

C. The forward voltage.

D. The back EMF.

3-10B5 In Figure 3B2, what is the voltage drop across R1?

A. 1.2 volts.   C. 3.7 volts.

B.   2.4 volts.   **D. 9 volts.**

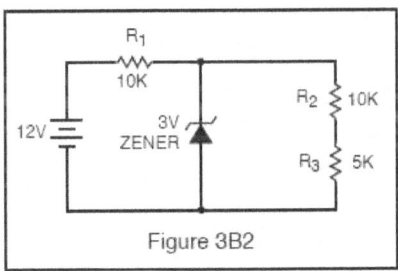

Figure 3B2

3-18B2 A 1-watt, 10-volt Zener diode with the following characteristics: $I_{min.} = 5$ mA; $I_{max.} = 95$ mA; and $Z = 8$ ohms, is to be used as part of a voltage regulator in a 20-V power supply. Approximately what size current-limiting resistor would be used to set its bias to the midpoint of its operating range?

A.   100 ohms.

B.   200 ohms.

C.   1 kilohms.

D.   2 kilohms.

3-19C4 What is the description of an optoisolator?

A.   An LED and a photosensitive device.

B.   A P-N junction that develops an excess positive charge when exposed to light.

C.   An LED and a capacitor.

D.   An LED and a lithium battery cell.

3-19C6 What is the description of an optocoupler?

A.   A resistor and a capacitor.

B.   Two light sources modulated onto a mirrored surface.

C,   An LED and a photosensitive device.

D.   An amplitude modulated beam encoder.

3-22C3 What device is usually used as a stable reference voltage in a linear voltage regulator?

A.   Zener diode.

B.   Tunnel diode.

C.   SCR.

D.   Varactor diode.

3-22C4 In a regulated power supply, what type of component will most likely be used to establish a reference voltage?

A.   Tunnel Diode.

B.   Battery.

C.   Pass Transistor.

D.   Zener Diode.

3-22C6 What is the range of voltage ratings available in Zener diodes?

A.   1.2 volts to 7 volts.

B.   2.4 volts to 200 volts and above.

C.   3 volts to 2000 volts.

D.   1.2 volts to 5.6 volts.

3-24C1 What is one common use for PIN diodes?

A. Constant current source.

B. RF switch.

C. Constant voltage source.

D. RF rectifier.

3-24C2 What is a common use of a hot-carrier diode?

A. Balanced inputs in SSB generation.

B. Variable capacitance in an automatic frequency control circuit.

C. Constant voltage reference in a power supply.

D. VHF and UHF mixers and detectors.

3-24C3 Structurally, what are the two main categories of semiconductor diodes?

A. Junction and point contact.

B. Electrolytic and junction.

C. Electrolytic and point contact.

D. Vacuum and point contact.

3-24C4 What special type of diode is capable of both amplification and oscillation?

A. Zener diodes.

B. Point contact diodes.

C. Tunnel diodes.

D. Junction diodes.

3-24C5 What type of semiconductor diode varies its internal capacitance as the voltage applied to its terminals varies?

A. Tunnel diode.

B. Varactor diode.

C. Silicon-controlled rectifier.

D. Zener diode.

3-24C6 What is the principal characteristic of a tunnel diode?

A. High forward resistance.

B. Very high PIV(peak inverse voltage).

C. Negative resistance region.

D. High forward current rating.

3-27C1 What type of bias is required for an LED to produce luminescence?

A. Reverse bias.

B. Forward bias.

C. Logic 0 (Lo) bias.

D. Logic 1 (Hi) bias.

3-27C2 What determines the visible color radiated by an LED junction?

A. The color of a lens in an eyepiece.

B. The amount of voltage across the device.

C. The amount of current through the device.

D. The materials used to construct the device.

3-27C3 What is the approximate operating current of a light-emitting diode?

A. 20 mA.

B. 5 mA.

C. 10 mA.

D. 400 mA.

3-27C4 What would be the maximum current to safely illuminate a LED?

A. 1 amp.

B. 1 microamp.

C. 500 milliamps.

D. 20 mA.

3-27C5 An LED facing a photodiode in a light-tight enclosure is commonly known as a/an:

A. Optoisolator.

B. Seven segment LED.

C. Optointerrupter.

D. Infra-red (IR) detector.

3-27C6 What circuit component must be connected in series to protect an LED?

A. Bypass capacitor to ground.

B. Electrolytic capacitor.

C. Series resistor.

D. Shunt coil in series.

3-28C1 What describes a diode junction that is forward biased?

A. It is a high impedance.

B. It conducts very little current.

C. It is a low impedance.

D. It is an open circuit.

3-32D2 Which lamps would be lit in the circuit shown in Figure 3D8?

A. 2, 3, 4, 5 and 6.

B. 5, 6, 8 and 9.

C. 2, 3, 4, 7 and 8.

D. 1, 3, 5, 7 and 8.

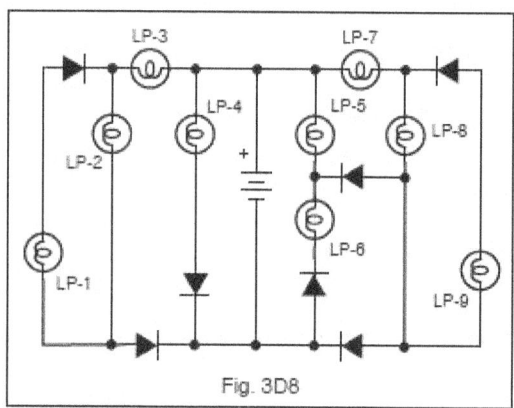

Fig. 3D8

3-32D6 With a pure AC signal input to the circuit shown in Figure

3D11, what output wave form would you expect to see on an oscilloscope display?

    A.   1

    B.   2

    C.   3

    D.   4

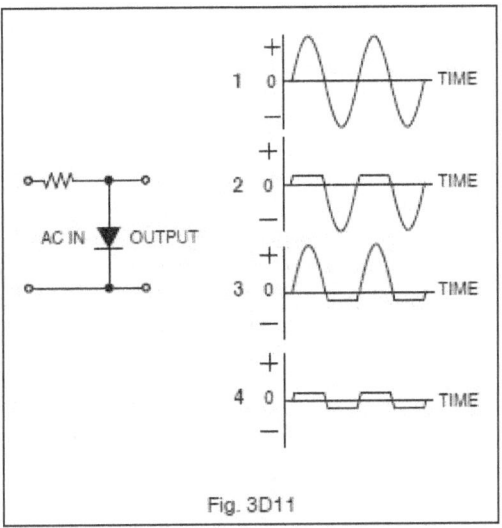

Fig. 3D11

3-48F3   What is the process of detection in a radio diode detector circuit?

    A.   Breakdown of the Zener voltage.

    B.   Mixing with noise in the transition region of the diode.

    C.   Rectification and filtering of RF.

    D.   The change of reactance in the diode with respect to frequency.

11: SEMICONDUCTORS

In addition to diodes, some other semiconductor devices are covered by the exam.

A **bipolar transistor** (in other words, a PNP transistor or an NPN transistor) has three terminals: Base, collector, and emitter. Bipolar transistors have two properties whose definitions you need to know for the exam. The "alpha" is defined as the collector current with respect to the emitter current. The "beta" is the collector current with respect to the base current. Remember that the collector current is always the numerator of this ratio, and the denominator for **beta** is the **base**. The **beta cutoff frequency** is defined as the frequency at shich the emitter current has decreased to 0.707 of maximum. Again, if you remember that emitter current is the numerator of the beta, this should be easy to remember, since the wrong answers refer to other currents. A transistor is fully saturated when the collector current is at its maximum value.

There is only one question regarding unijunction transistors. You need to know only that the three elements are base 1, base 2, and emitter. This is easy to remember, since the correct answer is the only one that uses names that are also used in bipolar junction transistors. The wrong answers all include the word "gate", which relates to FET's.

The two types of junction field-effect transistor (JFET) are N-channel and P-channel.

A MOSFET has a zener diode built into its gate to protect it from excessive voltage and static charges.

The difference you need to know between FET's and bipolar transistors is that bipolar transistors have a low input impedance, and FET's have a high input impedance. If you forget this, you might already know that FET's generally function based upon the input voltage, whereas bipolar junction transistors function based upon input current. Therefore, an FET should have a high input impedance, such as a voltmeter has, and the bipolar transistor should have a low input impedance, such as an ammeter has.

CMOS stands for complementary metal-oxide semiconductor. If you forget this, all you need to remember that CMOS is a type of semiconductor. It's not a system or a substrate, and there's no such thing as mica oxide. When handling CMOS devices (or FET's), you must remember that they are very susceptible to damage from static charges. When working with CMOS devices, it is often necessary to wear wrist straps to prevent static electricity. These should have less than 100,000 ohms resistance.

There are a number of questions on the exam regarding various photoelectric devices. A **photoconductive** device is one whose conductivity changes in the presence of light. When exposed to light, the conductivity of photoconductive material increases. In other words, the resistance decreases in the presence of light.

You need to know the following facts regarding voltage regulators: In a linear voltage regulator, the conduction of a control element is varied in proportion to either the line voltage or the load current. The internal components of a three-terminal regulator are a voltage reference, an error amplifier, sensing resistors and transistors, and a pass element.

In contrast to a linear voltage regulator, a switching voltage regulator switches the control device on or off to reduce the duty cycle in response to changes in the supply voltage and/or the load. A Silicon-controlled rectifier (SCR) is used to control voltage in applications such as a light dimmer. It reduces the average voltage by switching off for part of the cycle. The three terminals of an SCR are the anode, cathode, and gate. To distribute the power load of a circuit, two SCR's might be connected together in parallel, with reverse polarity.

A TRIAC is simply a combination of two SCR's connected back to back and sharing a common gate. Therefore, its terminals are gate, anode 1, and anode 2.

The test has a few questions discussing characteristics of various circuits. Generally, there is only a single question relating to each circuit. Therefore, these questions are best dealt with by memorization. If you already have an understanding of these circuits, then it probably won't be necessary to memorize the following facts. On the other hand, you'll get these questions right if you know only the following facts regarding some common transistor circuits:

A common-base amplifier has more voltage gain than a common-emitter. A common-emitter amplifier has more voltage gain than a common collector. An emitter-follower amplifier has more current gain than a common emtter or common base.

For the purposes of the exam, you can combine all of these facts into the following chart for easy memorization:

Emitter follower	MOST GAIN
Common base	SECOND HIGHEST GAIN
Common emitter	THIRD HIGHEST GAIN
Common collector	LOWEST GAIN

A transistor is operating in **saturation** when both junctions are forward biased. Finally, you need to know that for current to flow in an silicon NPN transistor's junction, the base must be 0.7 volts positive with respect to the emitter. If you remember only +.7, then you'll get this question right.

3-5A3 MOSFETs are manufactured with THIS protective device built into their gate to protect the device from static charges and excessive voltages:

A. Schottky diode.

B. Metal oxide varistor (MOV).

C. Zener diode.

D. Tunnel diode.

3-5A4 What are the two basic types of junction field-effect transistors?

A. N-channel and P-channel.

B. High power and low power.

C. MOSFET and GaAsFET.

D. Silicon FET and germanium FET.

3-5A5 A common emitter amplifier has:

A. Lower input impedance than a common base.

B. More voltage gain than a common collector.

C. Less current gain than a common base.

D. Less voltage gain than a common collector.

3-5A6 How does the input impedance of a field-effect transistor compare with that of a bipolar transistor?

A. An FET has high input impedance; a bipolar transistor has low input impedance.

B. One cannot compare input impedance without first knowing the supply voltage.

C. An FET has low input impedance; a bipolar transistor has high input impedance.

D. The input impedance of FETs and bipolar transistors is the same.

3-19C1 What happens to the conductivity of photoconductive material when light shines on it?

A. It increases.

B. It decreases.

C. It stays the same.

D. It becomes temperature dependent.

3-19C2 What is the photoconductive effect?

A. The conversion of photon energy to electromotive energy.

B. The increased conductivity of an illuminated semiconductor junction.

C. The conversion of electromotive energy to photon energy.

D. The decreased conductivity of an illuminated semiconductor junction.

3-19C3 What does the photoconductive effect in crystalline solids produce a noticeable change in?

A. The capacitance of the solid.

B. The inductance of the solid.

C. The specific gravity of the solid.

D. The resistance of the solid.

3-19C5 What happens to the conductivity of a photosensitive semiconductor junction when it is illuminated?

A. The junction resistance is unchanged.

B. The junction resistance decreases.

C. The junction resistance becomes temperature dependent.

D. The junction resistance increases

3-22C1 In a linear electronic voltage regulator:

    A.   The output is a ramp voltage.

    B.   The pass transistor switches from the "off" state to the "on"" state.

    C.   The control device is switched on or off, with the duty cycle proportional to the line or load conditions.

    D.   The conduction of a control element is varied in direct proportion to the line voltage or load current.

3-22C2 A switching electronic voltage regulator:

    A.   Varies the conduction of a control element in direct proportion to the line voltage or load current.

    B.   Provides more than one output voltage.

    C.   Switches the control device on or off, with the duty cycle proportional to the line or load conditions.

    D.   Gives a ramp voltage at its output.

3-22C5 A three-terminal regulator:

    A.   Supplies three voltages with variable current.

    B.   Supplies three voltages at a constant current.

    C.   Contains a voltage reference, error amplifier, sensing resistors and transistors, and a pass element.

    D.   Contains three error amplifiers and sensing transistors.

3-23C1 How might two similar SCRs be connected to safely distribute the power load of a circuit?

    A.   In series.

B.   In parallel, same polarity.

C.   In parallel, reverse polarity.

D.   In a combination series and parallel configuration.

3-23C2 What are the three terminals of an SCR?

A.   Anode, cathode, and gate.

B.   Gate, source, and sink.

C.   Base, collector, and emitter.

D.   Gate, base 1, and base 2.

3-23C3 Which of the following devices acts as two SCRs connected back to back, but facing in opposite directions and sharing a common gate?

A.   JFET.

B.   Dual-gate MOSFET.

C.   DIAC.

D.   TRIAC.

3-23C4 What is the transistor called that is fabricated as two complementary SCRs in parallel with a common gate terminal?

A.   TRIAC.

B.   Bilateral SCR.

C.   Unijunction transistor.

D.   Field effect transistor.

3-23C5 What are the three terminals of a TRIAC?

A. Emitter, base 1, and base 2.

B. Base, emitter, and collector.

C. Gate, source, and sink.

D. Gate, anode 1, and anode 2.

3-23C6 What circuit might contain a SCR?

A. Filament circuit of a tube radio receiver.

B. A light-dimming circuit.

C. Shunt across a transformer primary.

D. Bypass capacitor circuit to ground.

3-25C1 What is the meaning of the term "alpha" with regard to bipolar transistors? The change of:

A. Collector current with respect to base current.

B. Base current with respect to collector current.

C. Collector current with respect to gate current.

D. Collector current with respect to emitter current.

3-25C2 What are the three terminals of a bipolar transistor?

A. Cathode, plate and grid.

B. Base, collector and emitter.

C. Gate, source and sink.

D. Input, output and ground.

3-25C3 What is the meaning of the term "beta" with regard to bipolar transistors? The change of:

    A. Base current with respect to emitter current.

    B. Collector current with respect to emitter current.

    C. Collector current with respect to base current.

    D. Base current with respect to gate current.

3-25C4 What are the elements of a unijunction transistor?

    A. Base 1, base 2, and emitter.

    B. Gate, cathode, and anode.

    C. Gate, base 1, and base 2.

    D. Gate, source, and sink.

3-25C5 The beta cutoff frequency of a bipolar transistor is the frequency at which:

    A. Base current gain has increased to 0.707 of maximum.

    B. Emitter current gain has decreased to 0.707 of maximum.

    C. Collector current gain has decreased to 0.707.

    D. Gate current gain has decreased to 0.707.

3-25C6 What does it mean for a transistor to be fully saturated?

    A. The collector current is at its maximum value.

    B. The collector current is at its minimum value.

    C. The transistor's Alpha is at its maximum value.

    D. The transistor's Beta is at its maximum value.

3-26C1 A common base amplifier has:

A. More current gain than common emitter or common collector.

B. More voltage gain than common emitter or common collector.

C. More power gain than common emitter or common collector.

D. Highest input impedance of the three amplifier configurations.

3-26C2 What does it mean for a transistor to be cut off?

A. There is no base current.

B. The transistor is at its Class A operating point.

C. There is no current between emitter and collector.

D. There is maximum current between emitter and collector.

3-26C3 An emitter-follower amplifier has:

A. More voltage gain than common emitter or common base.

B. More power gain than common emitter or common base.

C. Lowest input impedance of the three amplifier configurations.

D. More current gain than common emitter or common base.

3-26C4 What conditions exists when a transistor is operating in saturation?

A. The base-emitter junction and collector-base junction are both forward biased.

B. The base-emitter junction and collector-base junction are both reverse biased.

C.   The base-emitter junction is reverse biased and the collector-base junction is forward biased.

D.   The base-emitter junction is forward biased and the collector-base junction is reverse biased.

3-26C5 For current to flow in an NPN silicon transistor's emitter-collector junction, the base must be:

A.   At least 0.4 volts positive with respect to the emitter.

B.   At a negative voltage with respect to the emitter.

C.   At least 0.7 volts positive with respect to the emitter.

D.   At least 0.7 volts negative with respect to the emitter.

3-28C2 Why are special precautions necessary in handling FET and CMOS devices?

A.   They have fragile leads that may break off.

B.   They are susceptible to damage from static charges.

C.   They have micro-welded semiconductor junctions that are susceptible to breakage.

D.   They are light sensitive.

3-28C3 What do the initials CMOS stand for?

A.   Common mode oscillating system.

B.   Complementary mica-oxide silicon.

C.   Complementary metal-oxide semiconductor.

D.   Complementary metal-oxide substrate.

3-79L1 When soldering or working with CMOS electronics products or equipment, a wrist strap:

A.   Must have less than 100,000 ohms of resistance to prevent static electricity.

B.   Cannot be used when repairing TTL devices.

C.   Must be grounded to a water pipe.

D.   Does not work.

12: WAVEFORMS

There are a number of questions on the exam regarding waveforms. Of course, the most common waveform we deal with is the sine wave. It is named after the trigonometric sine function, because if you draw a graph showing the sine of an angle, it will match the waveform. The definition you will encounter on the exam is probably unfamiliar. You can answer this question quite easily, because the other answers are clearly wrong. But on the test, a sine wave is defined as "a wave whose amplitude at any given instant can be represented by the projection of a point on a wheel rotating at a uniform speed."

On an oscilloscope, a sine wave would look like the following diagram of 60 Hz household electrical current. In this diagram, the horizontal axis shows time, and the vertical axis shows voltage:

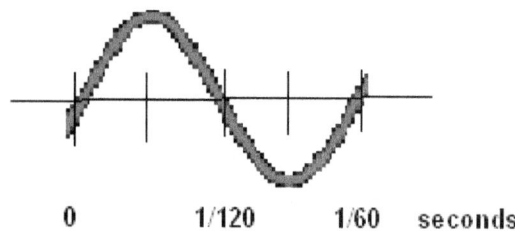

    0   1/120   1/60   seconds

As noted above, this graph is identical to the graph of the sine function.

The only difference is that the horizontal access is labeled in degrees.

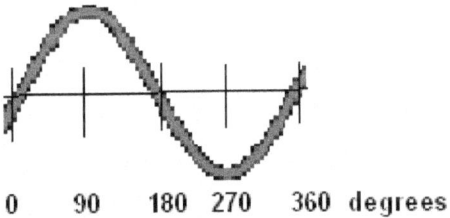

0   90   180 270   360 degrees

As you can see, the graph is identical, except how it is labeled. In the original drawing, one cycle was equal to 1/60 second. In the second drawing, one cycle was equal to 360 degrees. There is another way to label the graph, and that is using **radians**. 360 degrees equals one circle. If the radius of the circle is one unit, then the circumference of the circle is 2π. ($\pi = 3.14$). The following graph is identical, but instead of showing 0-360 degrees, it is marked 0 - 2π radians.

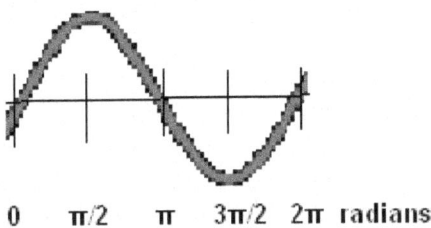

0    $\pi/2$ π $3\pi/2$ 2π radians

Depending on what we are doing, it might make our calculations easier depending on which way we label the diagram: time, degrees, or radians. The important thing to remember is that one complete cycle equals 360 degrees, which equals 2π radians.

There is a whole branch of mathematics called Fourier analysis, which is the study of how various waveforms can be represented as the sum or other waveforms. For the test, you need to know only the following facts:

1. A square wave is the sum of an infinite number of sine waves, starting with the fundamental frequency and all odd harmonics. (A square wave, as you probably know, is defined as one that "abruptly changes back and forth between two voltage levels and stays at these levels for equal amounts of time.")

2. A sawtooth wave is the sum of an infinite number of sine waves, starting with the fundamental frequency and including **all** harmonics. A sawtooth wave, as you probably know, looks like the blade of a saw. It is defined on the exam as one with "a rise time significantly faster than the fall time, or vice versa." You don't really need to memorize this definition. Of all the answers on the test, this is the only one that describes the edge of a saw.

The exam will ask you a number of questions about the voltage of a certain wave form at a certain point along its cycle. These questions give you a peak voltage, and ask you what the voltage is at a certain point, either in radians or degrees. If you have a calculator with the sine function, these problems are extremely easy. **But make sure that your calculator is set properly, for degrees or radians!** All you need to do is take the sine of the angle given, and multiply the result by the peak voltage that is given.

If you don't have your calculator with you, or if it doesn't do trigonometric functions, don't despair. On a piece of scratch paper, simply draw a sine wave. Label the horizontal access either in degrees or radians. If you're working in degrees, you'll start at zero, the positive peak will be 90, where it crosses the center line will be 180, the negative peak will be 270, and the end of the cycle will be 360. If the problem is in radians, then label these points 0, $\pi/2$ (1.57), π (3.14), $3\pi/2$ (4.61), and 2π (6.28). The highest point on the graph will be the peak voltage. The lowest point on the graph will be the negative peak voltage. Then, simply see what the approximate value is at the angle given in the question. Even if your graph is very poorly drawn, your answer will be close enough to answer the question.

We have been able to solve all of the problems on the exam by using the graphs shown above, which are what the mathematicians would call Cartesian coordinates. Some problems can be solved more readily by graphing things in polar coordinates. This, however, has not been necessary for the simple problems presented. However, there is one question which asks "using the polar coordinate system, what visual representation would you get of a voltage in a sinewave circuit?" The correct answer is "the plot shows the magnitude and phase angle." On this question, it is relatively easy to eliminate the wrong answers, so we won't spend further time discussing polar coordinates.

Finally, in our discussion of reactance, we discussed the concept of graphs "lagging" or "leading" one another. There's one additional question on the test, where you are merely asked to recognize in the following drawing that B is lagging A by 90 degrees:

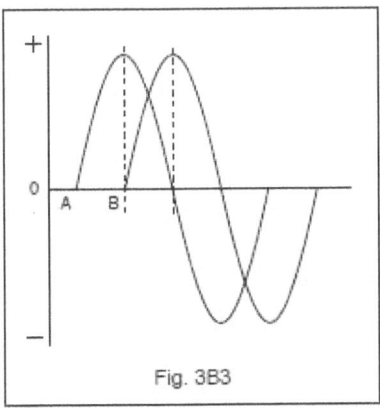

Fig. 3B3

3-7A1 What is a sine wave?

    A. A constant-voltage, varying-current wave.

    B. A wave whose amplitude at any given instant can be represented by the projection of a point on a wheel rotating at a uniform speed.

    C. A wave following the laws of the trigonometric tangent function.

    D. A wave whose polarity changes in a random manner.

3-7A2 How many degrees are there in one complete sine wave cycle?

A. 90 degrees.

B. 270 degrees.

C. 180 degrees.

D. 360 degrees.

3-7A3 What type of wave is made up of sine waves of the fundamental frequency and all the odd harmonics?

A. Square.

B. Sine.

C. Cosine.

D. Tangent.

3-7A4 What is the description of a square wave?

A. A wave with only 300 degrees in one cycle.

B. A wave whose periodic function is always negative.

C. A wave whose periodic function is always positive.

D. A wave that abruptly changes back and forth between two voltage levels and stays at these levels for equal amounts of time.

3-7A5 What type of wave is made up of sine waves at the fundamental frequency and all the harmonics?

A. Sawtooth wave.

B. Square wave.

C. Sine wave.

D. Cosine wave.

3-7A6 What type of wave is characterized by a rise time significantly faster than the fall time (or vice versa)?

    A. Cosine wave.

    B. Square wave.

    C. Sawtooth wave.

    D. Sine wave.

3-12B1 At pi/3 radians, what is the amplitude of a sine-wave having a peak value of 5 volts?

    A. -4.3 volts.   C. +2.5 volts.

    B. -2.5 volts.   **D. +4.3 volts.**

3-12B2 At 150 degrees, what is the amplitude of a sine-wave having a peak value of 5 volts?

    A. -4.3 volts.   **C. +2.5 volts.**

    B. -2.5 volts.   D. +4.3 volts.

3-12B3 At 240 degrees, what is the amplitude of a sine-wave having a peak value of 5 volts?

    A. -4.3 volts.   C. +2.5 volts.

    B. -2.5 volts.   D. +4.3 volts.

3-12B6 Determine the phase relationship between the two signals shown in Figure 3B3.

    A. A is lagging B by 90 degrees.

    B. B is lagging A by 90 degrees.

    C. A is leading B by 180 degrees.

D. B is leading A by 90 degrees.

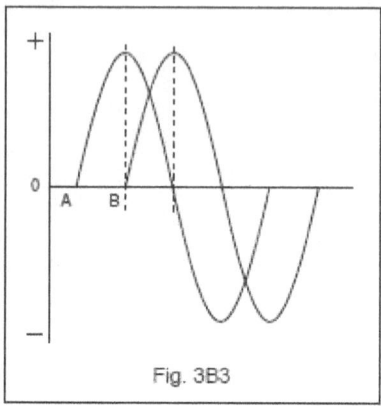

Fig. 3B3

3-17B6 Using the polar coordinate system, what visual representation would you get of a voltage in a sinewave circuit?

A. To show the reactance which is present.

B. To graphically represent the AC and DC component.

C. To display the data on an XY chart.

D. The plot shows the magnitude and phase angle.

13: AC ELECTRICITY, REACTANCE, IMPEDANCE, AND RESONANCE

In earlier chapters you learned about Ohm's Law and the concept of resistance. **Resistance**, as you recall, is the opposition to the flow of electricity in a circuit, and is measured in Ohms.

There is a similar concept, which only applies to alternating current (AC) circuits, and that concept is **impedance.** Like resistance, impedance is measured in Ohms, and Ohm's law still applies. As you recall, Ohm's law is expressed in the following three formulas:

Volts equals Ohms times Amps

Ohms equals Volts divided by Amps

Amps equals Volts divided by Ohms.

Impedance consists of both resistance and **reactance**. All three are measured in Ohms. Resistance applies to both AC and DC circuits. **Reactance** applies only to AC circuits. Just like resistance, reactance is the opposition to the flow of electricity. But reactance applies only to alternating current, and is caused by capacitance or inductance.

Impedance is the combination of both resistance and reactance, and is the total opposition to the flow of current in an AC circuit. However, you cannot simply add up the resistance and reactance to get the total impedance. Combining them uses a formula similar to the Pythagorean theorem, which you probably learned in high school. If resistance is one leg of a right triangle, and reactance is the other leg of the triangle, then the total impedance is equal to the hypotenuse of the triangle. For example, in the diagram below, we combine 3 ohms of resistance and 4 ohms of reactance. This gives us a total impedance of 5 ohms.

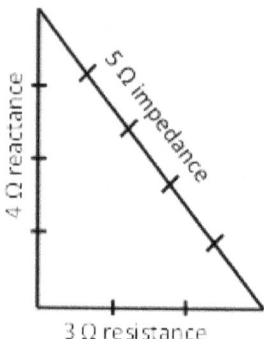

If you remember this triangle, you can always determine impedance without using any math. On a piece of scratch paper, draw a right triangle. The length of the horizontal piece would be equal to the resistance. You can use inches, centimeters, or any other unit you would like. The vertical height would be equal to the reactance. Complete the triangle, and the length of the longest side will be equal to the impedance.

We can also use graph paper to show how these figures are combined. On the following graph, the horizontal axis shows the resistance. The vertical axis shows the reactance. We plot the two numbers (3 ohms resistance and 4 ohms reactance), and plot this on the graph paper:

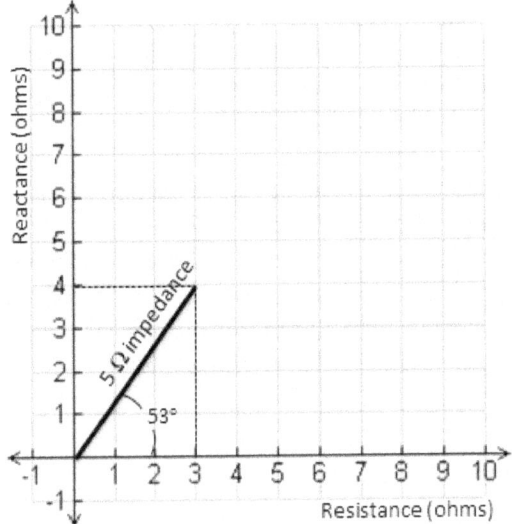

The line running to the point we put on the graph is 5 units long. This means that the total impedance is 5 ohms. You will also note that there is an angle at which this line runs. In this example, the angle is about 53 degrees.

In some cases, all we need to know is that the impedance is 5 ohms. However, in other cases, we want to know what the reactance and resistance are separately. There are two ways of writing this. Depending on what kind of problem we are doing, one might be more helpful than the other.

In some cases, it's important to know the angle. So for those problems, we'll want to know that the impedance was 5 ohms, and the angle is 53 degrees.

But in some cases, we'll need to do some math that involves both numbers. In that case, we use the letter "j" to refer to the reactance. So in this case, we would say that the impedance in ohms is:

$$3 + 4j$$

Most of the problems on the exam will involve writing the impedance this way. You might recall from high school algebra dealing with "imaginary numbers". The square root of negative one is an "imaginary number", and you probably remember from algebra that this number is called "i". For some reason, electrical engineers decided to use the letter "j" instead. But it's the same concept.

Don't be alarmed about these imaginary numbers. If you don't fully understand the math, you can still pass the test quite easily. All you really need to remember how impedance values are written. Whenever there is a mixture of resistance and reactance, it is written as resistance plus "j" times the reactance. We usually use the letter Z for impedance, and X for reactance. So we usually write the total impedance in the following way:

$$Z = R + jX$$

Reactance is caused by either inductance or capacitance. This means that every inductor and capacitor has a reactance that can be calculated or measured in ohms. This will vary depending on frequency, and will also vary depending upon the inductance or capacitance.

Reactance is not difficult to understand. Think for a minute about what a capacitor is. A capacitor, in its simplest form, consists of two metal plates separated by air or some other insulator. Look at the following diagram, and it should be clear that the light bulb will not light:

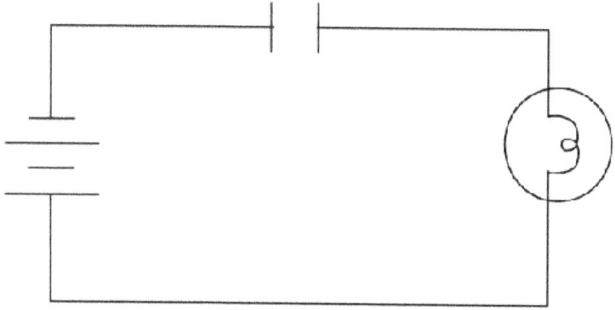

The light will not come on, because there is no path for the DC electricity. To put it another way, the resistance of the capacitor is infinity.

Now, look at the following diagram, where an inductor has replaced the capacitor:

As you recall, an inductor is nothing more than a coil of wire. From the

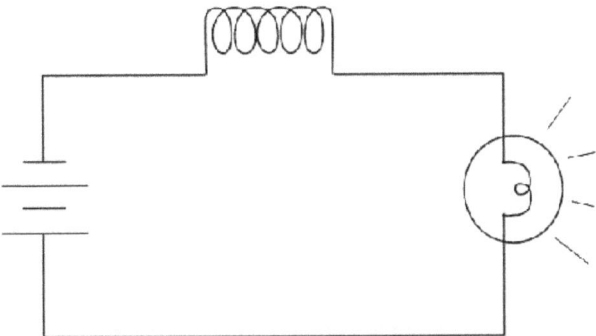

DC electricity's point of view, the lamp is connected to the battery by nothing more than an ordinary piece of wire. The DC electricity doesn't care that the wire is coiled up. Wire is wire, and the inductor has a resistance of zero ohms.

Now, let's replace the flashlight bulb with a 120 volt bulb, and replace the battery with AC electricity. As you know, standard household current is Alternating Current, and its frequency is 60 Hz (60 cycles per second). This means that the current changes direction 120 times per second. Every second, it is positive for 1/60 second, and it's negative for 1/60 second.

Here's what our light bulb and capacitor look like:

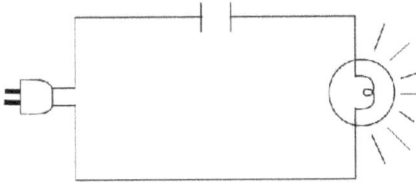

When you plug this in, there is no connection between the light bulb and the plug. Just like with the battery, the resistance of the capacitor is infinity. However, the electrons from the outlet want to get to the capacitor and charge up the plates. Remember, one side of the capacitor wants to charge up and become positive, and the other side wants to become negative. In order for this to happen, some of the electricity will flow through the bulb, and the bulb will light up for $1/120^{th}$ of a second.

$1/120^{th}$ of a second later, the current from the outlet will change direction. At this point, the charge on the positive side of the capacitor will need to flow to the negative side of the outlet. Again, some of this current will need to flow through the bulb, and the bulb will light up for another $1/120^{th}$ of a second.

This process keeps repeating itself, and every time the current changes direction, some of the current will flow through the bulb. Because the current is flowing through it, the bulb will light up. In other words, current is flowing through this circuit, even though there's a gap in the middle of the capacitor. The gap will block some of the flow, but it will not block all of it. In other words, it works just like resistance, and can be measured in ohms, just like a resistor. This "resistance" is called **capacitive reactance**.

As you remember, the bulb will not light with a battery and capacitor in series, because the DC electricity cannot flow through the capacitor. In other words, with DC, the capacitor acts just like an insulator, or a resistor resistance of infinity ohms. You can think of DC as having a frequency of zero. Remember, frequency is the number of times per second that the electricity changes direction. In the case of DC, it changes directions zero times per second, so the frequency is zero Hertz.

So we know that when the frequency is zero, the capacitive reactance is infinity. But we also know that when the frequency is higher than zero, the reactance goes down. It never gets to zero, but it keeps going down as we increase the frequency. For the test, you need to know that as frequency increases, capacitive reactance decreases.

In the case of an inductor, the process is just the opposite. As you recall, when working with DC (frequency zero), the inductor has absolutely no effect. It is just like a resistor of zero ohms. It's just a piece of wire, and the DC electricity doesn't care that it's coiled up.

But if we hook up a similar circuit using AC, the inductor will begin to have an effect. Take a look at the following diagram:

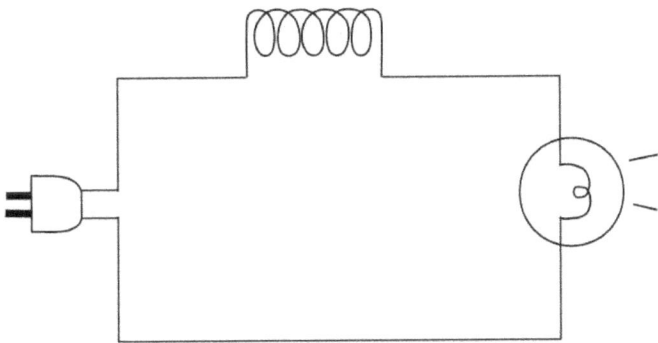

When the power is first hooked up, some of the energy is used to form a magnetic field around the coil. If you hold a compass near the coil, you would be able to see this field. Some of the energy is used to form this field.

But 1/120th second later, the current changes direction. As a result, the magnetic field changes polarity. Some of the electrical energy is used to take down the old magnetic field and put up the new field in the opposite direction. The energy to do this needs to come from somewhere, and as a result, the bulb will burn more dimly than it would otherwise. In other words, the inductor is acting just like a resistor, and this effect can be measured in Ohms. At a frequency of zero Hertz (DC), there is no effect, and this is zero ohms. As the frequency increases, the number of ohms increases. This property (the number of Ohms) is called **inductive reactance**. As the frequency increases, the inductive reactance increases.

The two types of reactance behave quite differently. (Capacitive reactance gets higher as the frequency gets closer to zero, and inductive reactance gets lower as the frequency gets closer to zero.) In fact, the two types of reactance generally "cancel out" if both are present. Because they behave differently, we often use different symbols. If we're referring to inductive reactance, we use the symbol X_L. If we're referring to capacitive reactance, we use the symbol X_C. For the exam, you should know the formulas for calculating each. For inductive reactance:

$$X_L = 2\pi fL$$

In this formula, f is the frequency in Hz, and L is the inductance in Henries. ($\pi = 3.14$).

For capacitive reactance, the formula is:

$$X_C = 1 / 2\pi f\,C$$

In this formula, f is the frequency in Hz, and C is the capacitance in Farads. **Think of capacitive reactance as being a negative number.** We'll see why in a moment.

You should memorize both of these formulas for the exam. But if you forget them when you're taking the exam, it's still possible to eliminate all or most of the wrong answers. Just think about what happens when the frequency is zero. In the capacitor, the number of Ohms is infinity, because a capacitor is just two metal plates that are not connected. In an inductor, the number of Ohms is zero, because an inductor is just a piece of wire, and the DC electricity does not care that the wire is coiled up.

There will be some questions on the exam that ask you to compute the total impedance, combining the reactance and reactance. These problems are quite simple. All you need to do is calculate the reactance using the formulas above. Then, on your scratch paper, do the following:

1. Write down the resistance that was given in the question. You will not change this number in any way.
2. If you are dealing with capacitive reactance, write down a minus sign.
3. If you were dealing with inductive reactance, write down a plus sign.
4. Write down the letter j.
5. Write down your answer from computing the reactance.

What you have written on your scratch paper should be the correct answer to the problem. Let's try this method with one of the test questions:

> In rectangular coordinates, what is the impedance of a network composed of a 0.1-microhenry inductor in series with a 30-ohm resistor, at 5 MHz?
>
> A. 30 -j3
>
> B. 3 +j30
>
> C. 3 -j30
>
> D. 30 +j3

For the first step, we simply write down 30, because that is the resistance. We're dealing with an inductor, so we write down a plus sign. Then, we write down the letter j, and we have:

$$30 + j \underline{\quad}.$$

In this problem, we didn't even need to calculate the reactance, because we've eliminated three of the answers! D is the only possible correct answer. In most cases, you will have eliminated all of the wrong answers before you even get to step 5! Try these simple steps on the exam questions. You'll discover that these problems, even though they look overwhelming, are quite easy, as long as you follow these steps.

There are also a number of questions that ask you to do the same kinds of problems, but give your answer in "polar coordinates". In other words, you will give a total number of ohms, and specify the angle. These problems also look difficult, but if we look at them carefully, we can figure out the correct answer, with minimal math.

First of all, on these problems, remember that if capacitance is involved, the angle will be a **negative number**. If inductance is involved, the angle will be a **positive number**. If you remember this, you will be able to eliminate most of the wrong answers. Let's look at the first question:

> What is the impedance of a network composed of a 100-picofarad capacitor in parallel with a 4000-ohm resistor, at 500 KIIz? Specify

your answer in polar coordinates.

A. 2490 ohms, /51.5 degrees

    B.   4000 ohms, /38.5 degrees

    C.   5112 ohms, /-38.5 degrees

    D.   2490 ohms, /-51.5 degrees

First of all, we're dealing with capacitance, so the angle will be a negative number. We can immediately eliminate A and B. We can pick out the right answer from the remaining two without doing any math. We have a 4000 ohm resistor in parallel with a capacitor, and we know that the capacitor has some reactance. We don't know what the reactance is, without calculating it. But it turns out we don't need to know the exact value. Remember, when you have resistors in parallel, the combined resistance is always less than either of the original resistors. And this is also true in the case of impedances. Therefore, we know the answer is going to be less than 4000. The only possible answer is D, and that is the correct answer.

Here's another easy one:

In polar coordinates, what is the impedance of a network comprised of a 400-ohm-reactance inductor in parallel with a 300-ohm resistor?

A. 240 ohms, /-36.9 degrees

B. 240 ohms, /36.9 degrees

C. 500 ohms, /53.1 degrees

D. 500 ohms, /-53.1 degrees

Again, we have two components in parallel. Therefore, we know that the combined value will be less than either of the original ones. The answer has to be less than 300, so we can eliminate C and D. Since we are dealing with inductance, the angle will be a positive number. Therefore, the answer must be B.

The next two problems can be done by drawing a diagram on a piece of scratch paper:

In polar coordinates, what is the impedance of a network composed of a 100-ohm-reactance inductor in series with a 100-ohm resistor?

Remember, reactance and resistance combine just like the two sides of a right triangle. So draw a right triangle that is 100 units along the bottom, and 100 units vertically. For example, you could make each side one inch long, and that represents 100 ohms. If you forgot your ruler, then just make one with another piece of scratch paper, or use your fingernail to measure your drawing. Then, finish the triangle and measure the long side with your ruler. It will be about 1.4 inches long, which means that the impedance is about 140 ohms. This is close to the right answer of 141. Since inductance was involved, the angle will be a positive number.

In polar coordinates, what is the impedance of a network composed of a 400-ohm-reactance capacitor in series with a 300-ohm resistor?

This problem is also easy to do with a simple drawing on a piece of scratch paper. Draw the horizontal line 3 inches long, to represent 300 ohms resistance. Then, draw the vertical line 4 inches long, to represent 400 ohms reactance. Finish the triangle, and measure the long side. It will be 5 inches long, meaning that the impedance is 500 ohms. Since we are dealing with capacitance, the angle will be a negative number. There is only one possible correct answer, 500 ohms, -53.1 degrees.

The last problem in this group is also quite simple:

In polar coordinates, what is the impedance of a network composed of a 300-ohm-reactance capacitor, a 600-ohm-reactance inductor, and a 400-ohm resistor, all connected in series?

A. **500 ohms, /37 degrees**

B. 400 ohms, /27 degrees

C. 300 ohms, /17 degrees

D. 200 ohms, /10 degrees

As mentioned above, the inductive reactance is positive, and the capacitive reactance is negative. Therefore, these two partially cancel out. We take the positive 600 ohms inductive reactance and the negative 300 ohms capacitive reactance. They partially cancel out, but not completely. So now, we are left with 300 ohms of reactance. (Even if we didn't know it was 300, we know that it is something.) So now, we have 400 ohms of resistance in series with 300 ohms of reactance. Therefore, we know that the answer is going to be greater than 400, and the only possible correct answer is A.

As we learned earlier, you calculate DC power by multiplying amps times volts. When we're dealing with AC, however, things get slightly more complicated. If any reactance is present, then the actual power is less than the result you get from this formula. But if you use the formula, the result you obtain is the **apparent power**. In other words, if you measure 120 AC volts and 2 AC amps, then the **apparent power** is 240 watts. If there is any reactance present, then the actual power might be less. The difference is the **reactive power**, which is defined on the exam as out-of-phase power associated with capacitors and inductors.

When the apparent power and the true power are equal (in other words, there is no reactance to worry about), then we say that the **power factor** is 1. Whenever there is reactance, the power factor is less than one. If there were reactance but no resistance, then the power factor would be zero.

A low power factor is caused by either a capacitor or inductor being in the circuit (unless both are present and they cancel each other out). Since a transformer consists of a coil of wire, it can add inductive reactance. Therefore, a capacitor is sometimes placed in series with a transformer to improve the power factor.

The reason for this is that when there is inductance or capacitance in a circuit, the current and voltage are **out of phase**. In other words, if you were to look at the current and voltage on an oscilloscope, the two waves would be offset. If there is only inductance, then the current lags the voltage. If there is only capacitance, then the voltage lags the current. To remember this, think of "Eli the Iceman". Eli has been delivering ice to the electronics industry for years, and his name helps us remember the relationship. Remember, E stands for voltage, L stands for inductance, I stands for current, and C stands for capacitance. So we have ELI, which reminds us that if L (inductance) is present, then E (voltage) comes before I (current). And we have ICE, which reminds us that if we have C (capacitance) present, then I (current) comes before E (voltage).

Since the voltage and current are happening at different times, this is why we don't get all of the power we were expecting. The power factor goes down. DC power eventually turns to heat. This reactive power never does so. Instead, it alternates between magnetic and electric fields.

Fortunately, there are only three questions on the test that ask you to calculate the power factor. Since there are only three, it's probably easiest to memorize the answer:

Question: What does the power factor equal in an R-L circuit having a 60 degree phase angle between the voltage and the current?

Answer: 0.5

Question: What does the power factor equal in an R-L circuit having a 45 degree phase angle between the voltage and the current?

Answer: 0.707

Question: What does the power factor equal in an R-L circuit having a 30 degree phase angle between the voltage and the current?

Answer: 0.866

Even if you forget these answers, you can eliminate some wrong answers by remembering that a power factor is always a number between 0 and 1. As the angle gets closer to zero, the power factor gets closer to 1.

There are questions on the test about **impedance matching**, but the questions do not go into detail. The main thing that you need to know is that when the impedance of a source and a load are matched (in other words, the same number of ohms), then there will be the maximum transfer of power.

One way to match the impedance is to insert an LC network between the source and the load. An "LC network" is simply an inductor (coil) and a capacitor. This might also be called a "Pi network", because the coil and capacitors might be arranged to look like the Greek letter Pi. A pi network can also consist of two inductors and one capacitor, again shaped like the letter pi.

An L-network consists of an inductor and a capacitor, connected together in the shape of the letter L.

The other type of matching network is the pi-L network, which is a combination of the two. It consists of two inductors and two capacitors. The Pi-L network provides the greatest amount of harmonic suppression.

Another method would be an impedance matching transformer. That is covered in more detail in the chapter on transformers.

Resonance is reached at the frequency where the inductive reactance and the capacitive reactance are equal. As we have seen above, they generally cancel out at this point. Because a number of interesting properties take place at resonance, a tunable L-C circuit is often used to adjust frequency of a circuit. Since the inductive and capacitive reactance cancel out, all that is left is whatever resistance is present. Therefore, as two of the questions note, the impedance of a series or parallel R-L-C circuit at resonance is approximately equal to the resistance.

At resonance, because they are canceling out, the individual voltages across the reactances can be greater than the applied voltage.

In an R-L-C circuit, the current flow is maximum at resonance. Also, at resonance, the voltage and current are in phase.

If you are matching an antenna to a transmitter, one method of impedance matching might even include using a transmission line of a particular length.

3-1A1 The product of the readings of an AC voltmeter and AC ammeter is called:

A. Apparent power.

B. True power.

C. Power factor.

D. Current power.

3-1A5 What formula would determine the inductive reactance of a coil if frequency and coil inductance are known?

A. $X_L = \pi f L$

B. $X_L = 2\pi f L$

C. $X_L = 1 / 2\pi f C$

D. $X_L = 1 / R2 + X2$

3-1A6 What is the term for the out-of-phase power associated with inductors and capacitors?

A. Effective power.

B. True power.

C. Peak envelope power.

D. Reactive power.

3-8A2 What happens to reactive power in a circuit that has both inductors and capacitors?

A. It is dissipated as heat in the circuit.

B. It alternates between magnetic and electric fields and is not dissipated.

C. It is dissipated as inductive and capacitive fields.

D. It is dissipated as kinetic energy within the circuit.

3-8A5 How do you compute true power (power dissipated in the circuit) in a circuit where AC voltage and current are out of phase?

A. Multiply RMS voltage times RMS current.

B. Subtract apparent power from the power factor.

C. Divide apparent power by the power factor.

D. Multiply apparent power times the power factor.

3-8A6 Assuming a power source to have a fixed value of internal resistance, maximum power will be transferred to the load when:

A. The load impedance is greater than the source impedance.

B. The load impedance equals the internal impedance of the source.

C. The load impedance is less than the source impedance.

D. The fixed values of internal impedance are not relative to the power source.

3-13B1 What does the power factor equal in an R-L circuit having a 60 degree phase angle between the voltage and the current?

A. 0.414

B. 0.866

C. 0.5

D. 1.73

3-13B4 In a circuit where the AC voltage and current are out of phase, how can the true power be determined?

A. **By multiplying the apparent power times the power factor.**

B. By subtracting the apparent power from the power factor.

C. By dividing the apparent power by the power factor.

D. By multiplying the RMS voltage times the RMS current.

3-13B5 What does the power factor equal in an R-L circuit having a 45 degree phase angle between the voltage and the current?

A. 0.866

B. 1.0

C. 0.5

D. **0.707**

3-13B6 What does the power factor equal in an R-L circuit having a 30 degree phase angle between the voltage and the current?

A. 1.73

B. **0.866**

C. 0.5

D. 0.577

3-16B1 What is the impedance of a network composed of a 0.1-microhenry inductor in series with a 20-ohm resistor, at 30 MHz? Specify your answer in rectangular coordinates.

A. 20 -j19

B. 19 +j20

C. **20 +j19**

D. 19 -j20

3-16B2 In rectangular coordinates, what is the impedance of a network composed of a 0.1-microhenry inductor in series with a 30-ohm resistor, at 5 MHz?

A. 30 -j3

B. 3 +j30

C. 3 -j30

D. 30 +j3

3-16B3 In rectangular coordinates, what is the impedance of a network composed of a 10-microhenry inductor in series with a 40-ohm resistor, at 500 MHz?

A. 40 +j31400

B. 40 -j31400

C. 31400 +j40

D. 31400 -j40

3-16B4 In rectangular coordinates, what is the impedance of a network composed of a 1.0-millihenry inductor in series with a 200-ohm resistor, at 30 kHz?

A. 200 - j188

B. 200 + j188

C. 188 + j200

D. 188 - j200

3-16B5 In rectangular coordinates, what is the impedance of a network composed of a 0.01-microfarad capacitor in parallel with a 300-ohm resistor, at 50 kHz?

A. 150 - j159

B. 150 + j159

C. 159 - j150

D. 159 + j150

3-16B6 In rectangular coordinates, what is the impedance of a network composed of a 0.001-microfarad capacitor in series with a 400-ohm resistor, at 500 kHz?

A. 318 - j400

B. 400 + j318

C. 318 + j400

D. 400 - j318

3-17B1 What is the impedance of a network composed of a 100-picofarad capacitor in parallel with a 4000-ohm resistor, at 500 KHz? Specify your answer in polar coordinates.

A. 2490 ohms, /51.5 degrees

B. 4000 ohms, /38.5 degrees

C. 5112 ohms, /-38.5 degrees

D. 2490 ohms, /-51.5 degrees

3-17B2 In polar coordinates, what is the impedance of a network composed of a 100-ohm-reactance inductor in series with a 100-ohm resistor?

A. 121 ohms, /35 degrees

B. 141 ohms, /45 degrees

C. 161 ohms, /55 degrees

D. 181 ohms, /65 degrees

3-17B3 In polar coordinates, what is the impedance of a network composed of a 400-ohm-reactance capacitor in series with a 300-ohm resistor?

A. 240 ohms, /36.9 degrees

B. 240 ohms, /-36.9 degrees

C. 500 ohms, /-53.1 degrees

D. 500 ohms, /53.1 degrees

3-17B4 In polar coordinates, what is the impedance of a network composed of a 300-ohm-reactance capacitor, a 600-ohm-reactance inductor, and a 400-ohm resistor, all connected in series?

A. 500 ohms, /37 degrees

B. 400 ohms, /27 degrees

C. 300 ohms, /17 degrees

D. 200 ohms, /10 degrees

3-17B5 In polar coordinates, what is the impedance of a network comprised of a 400-ohm-reactance inductor in parallel with a 300-ohm resistor?

A. 240 ohms, /-36.9 degrees

B. 240 ohms, /36.9 degrees

C.   500 ohms, /53.1 degrees

D.   500 ohms, /-53.1 degrees

3-18B1 What is the magnitude of the impedance of a series AC circuit having a resistance of 6 ohms, an inductive reactance of 17 ohms, and zero capacitive reactance?

A.   6.6 ohms.

B.   11 ohms.

C.   18 ohms.

D.   23 ohms.

3-21C1 A capacitor is sometimes placed in series with the primary of a power transformer to:

A.   Improve the power factor.

B.   Improve output voltage regulation.

C.   Rectify the primary windings.

D.   None of these.

3-29D1 What is the approximate magnitude of the impedance of a parallel R-L-C circuit at resonance?

A.   Approximately equal to the circuit resistance.

B.   Approximately equal to X_L.

C.   Low, as compared to the circuit resistance.

D.   Approximately equal to X_C.

3-29D2 What is the approximate magnitude of the impedance of a series R-L-C circuit at resonance?

A. High, as compared to the circuit resistance.

B. Approximately equal to the circuit resistance.

C. Approximately equal to X_L.

D. Approximately equal to X_C.

3-29D3 How could voltage be greater across reactances in series than the applied voltage?

A. Resistance.

B. Conductance.

C. Capacitance.

D. Resonance.

3-29D4 What is the characteristic of the current flow in a series R-L-C circuit at resonance?

A. Maximum.

B. Minimum.

C. DC.

D. Zero.

3-29D5 What is the characteristic of the current flow within the parallel elements in a parallel R-L-C circuit at resonance?

A. Minimum.

B. Maximum.

C. DC.

D. Zero.

3-29D6 What is the relationship between current through a resonant circuit and the voltage across the circuit?

    A. The current and voltage are 180 degrees out of phase.

    B. The current leads the voltage by 90 degrees.

    C. The voltage and current are in phase.

    D. The voltage leads the current by 90 degrees.

3-54G1 What is an L-network?

    A. A low power Wi-Fi RF network connection.

    B. A network consisting of an inductor and a capacitor.

    C. A "lossy" network.

    D. A network formed by joining two low pass filters.

3-54G2 What is a pi-network?

    A. A network consisting of a capacitor, resistor and inductor.

    B. The Phase inversion stage.

    C. An enhanced token ring network.

    D. A network consisting of one inductor and two capacitors or two inductors and one capacitor.

3-54G3 What is the resonant frequency in an electrical circuit?

    A. The frequency at which capacitive reactance equals inductive reactance.

    B. The highest frequency that will pass current.

    C. The lowest frequency that will pass current.

D.   The frequency at which power factor is at a minimum.

3-54G4 Which three network types are commonly used to match an amplifying device to a transmission line?

A.   Pi-C network, pi network and T network.

B.   T network, M network and Z network.

C.   L network, pi network and pi-L network.

D.   L network, pi network and C network.

3-54G5 What is a pi-L network?

A.   A Phase Inverter Load network.

B.   A network consisting of two inductors and two capacitors.

C.   A network with only three discrete parts.

D.   A matching network in which all components are isolated from ground.

3-54G6 Which network provides the greatest harmonic suppression?

A.   L network.

B.   Pi network.

C.   Pi-L network.

D.   Inverse L network.

14: TIME CONSTANTS

Whenever you pair up a resistor and a capacitor, or a resistor and an inductor, there is a time constant associated with the combination. In the case of a capacitor, the time constant (measured in seconds) is equal to the resistance times the capacitance:

$T = RC$

In the case of an inductor, it is:

$T = R / L$

In these formulas, C and L are measured in Farads and Henries. So if the problem calls for microfarads or millihenries, don't forget to divide the number given by one million or one thousand.

Some of the problems indicate that there are capacitors or resistors in series and/or in parallel. Before you do these problems, combine the resistances or capacitances so that you are dealing with a single resistor and a single capacitor. Then, calculate the time constant.

The time constant is the length of time that it takes the capacitor to charge to 63.2% of the supply voltage. In the case of an inductor, it's the length of time that it takes the current to build up to 63.2% of the maximum value.

A capacitor does not charge up immediately. It takes a certain amount of time, and the time constant tells us how long. Mathematically, the charge keeps increasing exponentially. For the purpose of the exam, we don't need to know the exponential formula, as long as we know that during each time constant, the voltage (or current, in the case of an inductor) gets 63.2% closer to the maximum than it was at the beginning of that time period.

For example, let's assume that the power supply voltage is 10 volts, and the time constant of a resistor and capacitor works out to one second. This means that after one second, the voltage across the capacitor is 6.32 volts. In other words, it's still 3.68 volts away from the maximum. During the next time constant period, the charge will increase by an additional 63.2% of the remaining value. 63.2% x 3.68 = 2.33. So after two seconds, the voltage would be 6.32 + 2.33 = 8.65 volts. There are still 1.35 volts left to go, so during the next time constant, the voltage would increase by 63.2% of 1.35 volts.

The same process works in reverse: The time constant applies to a capacitor discharging. If the capacitor starts out at 10 volts, then after one second, the voltage has decreased by 63.2%, so the new voltage is 3.68 volts. During the second second, the capacitor loses 63.2% of that amount, and the voltage at the end of that period is 1.35 volts.

The exam has one question asking you to identify the wave form of a capacitor alternately charging and then discharging. The correct answer is number 3 below. As you can see, when the capacitor is charging, it starts of fast, and then the rate of increase slows down. Similarly, when it discharges, it starts out fast, but then slows down. Number 2 is the wrong answer, because it starts charging slowly and then speeds up, and also starts discharging slowly and then speeds up. Numbers 1 and 4 are obviously wrong answers.

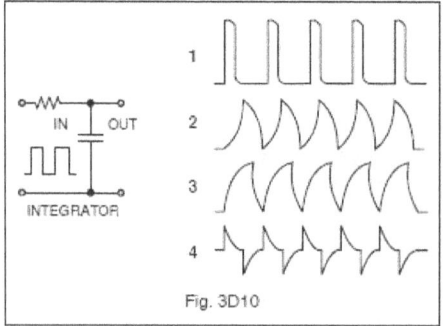

Fig. 3D10

3-14B1 What is the term for the time required for the capacitor in an RC circuit to be charged to 63.2% of the supply voltage?

  A. An exponential rate of one.

  B. One time constant.

  C. One exponential period.

  D. A time factor of one.

3-14B2 What is the meaning of the term "time constant of an RC circuit"? The time required to charge the capacitor in the circuit to:

  A. 23.7% of the supply voltage.

  B. 36.8% of the supply voltage.

  C. 57.3% of the supply voltage.

  D. 63.2% of the supply voltage.

3-14B3 What is the term for the time required for the current in an RL circuit to build up to 63.2% of the maximum value?

  A. One time constant.

  B. An exponential period of one.

  C. A time factor of one.

  D. One exponential rate.

3-14B4 What is the meaning of the term "time constant of an RL circuit"? The time required for the:

  A. Current in the circuit to build up to 36.8% of the maximum value.

  B. Voltage in the circuit to build up to 63.2% of the maximum

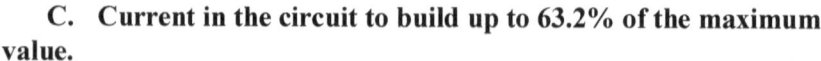

value.

C. Current in the circuit to build up to 63.2% of the maximum value.

D. Voltage in the circuit to build up to 36.8% of the maximum value.

3-14B5 After two time constants, the capacitor in an RC circuit is charged to what percentage of the supply voltage?

A. 36.8 %

B. 63.2 %

C. 86.5 %

D. 95 %

3-14B6 After two time constants, the capacitor in an RC circuit is discharged to what percentage of the starting voltage?

A. 86.5 %

B. 13.5 %

C. 63.2 %

D. 36.8 %

3-15B1 What is the time constant of a circuit having two 220-microfarad capacitors and two 1-megohm resistors all in parallel?

A. 22 seconds.

B. 44 seconds.

C. 440 seconds.

D. 220 seconds.

3-15B2 What is the time constant of a circuit having two 100-microfarad capacitors and two 470-kilohm resistors all in series?

   A. 470 seconds.

   B. 47 seconds.

   C. 4.7 seconds.

   D. 0.47 seconds.

3-15B3 What is the time constant of a circuit having a 100-microfarad capacitor and a 470-kilohm resistor in series?

   A. 4700 seconds.

   B. 470 seconds.

   C. 47 seconds.

   D. 0.47 seconds.

3-15B4 What is the time constant of a circuit having a 220-microfarad capacitor and a 1-megohm resistor in parallel?

   A. 220 seconds.

   B. 22 seconds.

   C.   2.2 seconds.

   D. 0.22 seconds.

3-15B5 What is the time constant of a circuit having two 100-microfarad capacitors and two 470-kilohm resistors all in parallel?

   A. 470 seconds.

   B. 47 seconds.

C. 4.7 seconds.

D. 0.47 seconds.

3-15B6 What is the time constant of a circuit having two 220-microfarad capacitors and two 1-megohm resistors all in series?

A. 220 seconds.

B. 55 seconds.

C. 110 seconds.

D. 440 seconds.

3-32D5 In Figure 3D10 with a square wave input what would be the output?

A. 1

B. 2

C. 3

D. 4

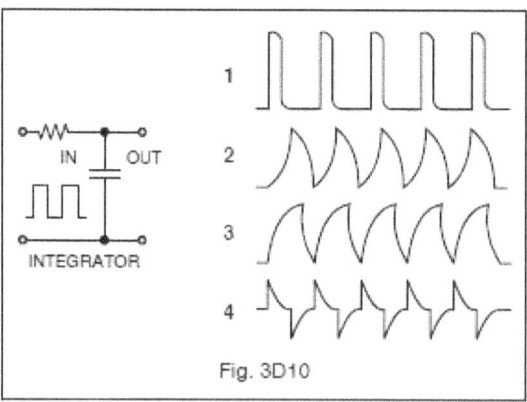

Fig. 3D10

15: TRANSFORMERS

The test covers a number of questions about transformers.

A **transformer** is simply two coils of wire placed next to each other. When AC power is connected to one of the coils, this causes a magnetic field. The magnetic field can cause an electric current to flow in the other coil. This phenomenon is called **mutual inductance**. The coil that is actually hooked up to the power source is called the **primary.** The other coil is called the **secondary**.

Here is the schematic symbol for a transformer:

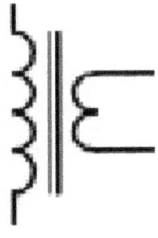

TRANSFORMER

One common use of a transformer is to change voltage. When one AC voltage is hooked up to the primary, another voltage will appear on the secondary. This depends on the ratio of the number of winding turns. For example, if there are 100 turns on the primary, and 1000 turns in the secondary, then the voltage will go up by a factor of ten: If the primary is hooked up to 120 volts, then the secondary voltage would be 1200 volts.

If the primary was 1000 turns and the secondary was 100 turns, then the voltage would go down by a factor of 10. In other words, if the primary is hooked up to 120 volts, then the secondary voltage would be 12 volts.

To put this another way, the ratio of voltages is exactly the same as the ratio of turns. If the secondary of the transformer is twice as big as the primary, then the voltage will be twice as big. If the secondary has three times the number of turns as the primary, then the secondary voltage will be three times as high.

The questions on the test are relatively simple, and can be figured out if you read them carefully and remember that the ratio of voltage is the same as the ratio of turns. It's probably not very helpful to memorize a formula. If you remember what is going on, you can easily figure out these questions.

As we learned earlier, a transformer can also be used to match impedance. The math is only slightly more complicated, but still not very difficult. In the case of impedance, you need to remember that the turns ratio is the square root of the impedance ratio. Or to put it another way, the turns ratio multiplied by itself equals the impedance ratio.

For example, if the primary has 100 turns, and the secondary has 500, this is a turns ratio of 5. Therefore, the impedance would change by a factor 25, because 5 x 5 = 25.

One fact about transformers is covered in the section of the test regarding aviation equipment. Many aircraft have power systems that operate at 400 Hz, rather than 60 Hz. This is because magnetic devices such as transformers can be smaller and lighter at a higher frequency.

3-18B4 What turns ratio does a transformer need in order to match a source impedance of 500 ohms to a load of 10 ohms?

A. 7.1 to 1.

B. 14.2 to 1.

C. 50 to 1.

D. None of these.

3-21C2 A transformer used to step up its input voltage must have:

A. More turns of wire on its primary than on its secondary.

B. More turns of wire on its secondary than on its primary.

C. Equal number of primary and secondary turns of wire.

D. None of the above statements are correct.

3-21C3 A transformer primary of 2250 turns connected to 120 VAC will develop what voltage across a 500-turn secondary?

A. 26.7 volts.

B. 2300 volts.

C. 1500 volts.

D. 5.9 volts.

3-21C5 A power transformer has a single primary winding and three secondary windings producing 5.0 volts, 12.6 volts, and 150 volts. Assuming similar wire sizes, which of the three secondary windings will have the highest measured DC resistance?

A. The 12.6 volt winding.

B. The 150 volt winding.

C. The 5.0 volt winding.

D. All will have equal resistance values.

3-21C6 A power transformer has a primary winding of 200 turns of #24 wire and a secondary winding consisting of 500 turns of the same size wire. When 20 volts are applied to the primary winding, the expected secondary voltage will be:

A.   500 volts.

B.   25 volts.

C.   10 volts.

D.   50 volts.

3-73K1 Some aircraft and avionics equipment operates with a prime power line frequency of 400 Hz. What is the principle advantage of a higher line frequency?

   A.   400 Hz power supplies draw less current than 60 Hz supplies allowing more current available for

other systems on the aircraft.

   B.   A 400 Hz power supply generates less heat and operates much more efficiently than a 60 Hz power supply.

   C.   The magnetic devices in a 400 Hz power supply such as transformers, chokes and filters are smaller and lighter than those used in 60 Hz power supplies.

   D.   400 Hz power supplies are much less expensive to produce than power supplies with lower line frequencies.

16: DECIBELS

One unit that you need to be aware of for the test is the **decibel (dB)**. A decibel is a unit used to measure the ratio of two different quantities, usually power. The main fact that you need to know is that anytime the power doubles, this is an increase of 3 dB. So if the power goes from one watt to two watts, this is an increase of 3 dB. If it doubles twice (in other words, it goes from the original value to four times as much), this would be a total of 6 dB.

For example, if we start with 10 watts, and we want to know how many dB increase it would be if it went up to 40 watts, here is how you would do the problem. First, you increase it to 20 watts. When this happens, it went up 3 dB, because it doubled. (Remember, that's the one fact that you should memorize—doubling is the same as a 3 dB increase). Then, you increase it again to 40 watts. This increased it by 3 more dB, for a total of 6 dB.

If it went up by a factor of 8 (for example, it went from 100 watts to 800 watts), this would be about 9 dB, because we had to double it three times (first from 100 to 200, then from 200 to 400, and then from 400 to 800).

You should be able to solve most of the problems this way. But another fact that will make some of the problems easier is that an increase by a factor of 10 is equal to exactly 10 dB. For example, if we went from 100 watts to 1000 watts, this would be 10 dB. (Note that this is similar to the last problem—an increase from 100 watts to 800 watts is about 9 dB, so it makes sense that this answer will be slightly higher, which it is.)

One item that is usually measured in dB is the amount of loss of a particular kind of feed line. It is usually measured in dB per hundred feet. So if you have a 50 foot piece of cable with a loss of 6 dB per hundred feet, then the loss would be about 3 dB. In other words, the signal at the output would be about half as strong as the input.

Finally, if you plan to use a calculator on the test and want to memorize the formula that can be used for any of the decibel problems, here is the formula.

$$dB = 10 \log (P1/P2)$$

P1 and P2 are the two values that you are comparing, and the log is the base 10 logarithm on your calculator. But if you don't want to use this formula, remember, you can use two rules to solve most of the problems on the test: First, 3 dB is the same as doubling, and 10 dB is the same as increasing by a factor of 10.

There's one antenna question that can be answered by using nothing but common sense and an elementary knowledge of decibels: "How does the gain of a parabolic dish antenna change when the operating frequency is doubled." If the frequency is doubled, this means that the wavelength is cut in half. This means that the antenna is now, effectively, twice as large, in terms of length. This means that it is not four times the effective area. Therefore, it stands to reason that the received signal will be four times as strong, which, as we saw above, means an increase of 6 dB.

3-64J5 An antenna radiates a primary signal of 500 watts output. If there is a 2nd harmonic output of 0.5 watt, what attenuation of the 2nd harmonic has occurred?

A. 10 dB.

B. 30 dB.

C. 40 dB.

D. 50 dB.

3-64J6 There is an improper impedance match between a 30 watt transmitter and the antenna, with 5 watts reflected. How much power is actually radiated?

    A. 35 watts.

    B. 30 watts.

    C. 25 watts.

    D. 20 watts.

3-66J6 If a transmission line has a power loss of 6 dB per 100 feet, what is the power at the feed point to the antenna at the end of a 200 foot transmission line fed by a 100 watt transmitter?

    A. 70 watts.

    B. 50 watts.

    C. 25 watts.

    D. 6 watts.

3-67J1 What is the effective radiated power of a repeater with 50 watts transmitter power output, 4 dB feedline loss, 3 dB duplexer and circulator loss, and 6 dB antenna gain?

    A. 158 watts.

    B. 39.7 watts.

    C. 251 watts.

    D. 69.9 watts.

3-67J2 What is the effective radiated power of a repeater with 75 watts transmitter power output, 4 dB feedline loss, 3 dB duplexer and circulator loss, and 10 dB antenna gain?

A. 600 watts.

B. 75 watts.

C. 18.75 watts.

D. 150 watts.

3-67J3 What is the effective radiated power of a repeater with 75 watts transmitter power output, 5 dB feedline loss, 4 dB duplexer and circulator loss, and 6 dB antenna gain?

A. 37.6 watts.

B. 237 watts.

C. 150 watts.

D. 23.7 watts.

3-67J4 What is the effective radiated power of a repeater with 100 watts transmitter power output, 4 dB feedline loss, 3 dB duplexer and circulator loss, and 7 dB antenna gain?

A. 631 watts.

B. 400 watts.

C. 25 watts.

D. 100 watts.

3-67J5 What is the effective radiated power of a repeater with 100 watts transmitter power output, 5 dB feedline loss, 4 dB duplexer and circulator loss, and 10 dB antenna gain?

A. 126 watts.

B. 800 watts.

C.   12.5 watts.

D.   1260 watts.

3-67J6 What is the effective radiated power of a repeater with 50 watts transmitter power output, 5 dB feedline loss, 4 dB duplexer and circulator loss, and 7 dB antenna gain?

A.   300 watts.

B.   315 watts.

C.   31.5 watts.

D.   69.9 watts.

3-93O1 How does the gain of a parabolic dish antenna change when the operating frequency is doubled?

A.   Gain does not change.

B.   Gain is multiplied by 0.707.

C.   Gain increases 6 dB.

D.   Gain increases 3 dB.

17: OSCILLATORS

An **oscillator** is a circuit that produces any type of alternating current, including AF or RF. An oscillator is essentially an amplifier with part of the output been fed back to the input. In order of the circuit to oscillate, there must be sufficient positive feedback. The frequency of oscillation depends on the amount of time it takes for the signal to feed back. The three major types of oscillator found in radio equipment are Colpitts, Hartley, and Pierce.

The **Colpitts** oscillator is used most often as a VFO (variable frequency oscillator) because it is very stable.

A **Pierce** oscillator contains a quartz crystal which sets the frequency of oscillation. The crystal vibrates as a result of the piezoelectric effect, which is a mechanical vibration of the crystal caused by application of voltage.

Finally, another type of oscillator is the VCO (voltage controlled oscillator). It is typically used in frequency synthesizers. A phase-locked loop (PLL) compares the output of the VCO to a frequency standard, and uses that output to adjust the internal capacitance of a varactor diode. One of the questions words the definition a different way: A PLL circuit is a servo loop containing a phase detector, a low-pass filter, and a VCO. A PLL synthesizer has the disadvantage of producing broadband noise.

The other type of frequency synthesizer is a direct digital synthesizer (DDS), which uses a phase comparator, look-up table, digital to analog converter (DAC), and a low-pass antialias filter. The disadvantage of a DDS is that it produces spurious signals (spurs) at discrete frequencies.

It should be noted that the abbreviation "DAC" stands for digital-to-analog converter, and "ADC" stands for analog-to-digital converter.

3-28C4   What is the piezoelectric effect?

A.   Mechanical vibration of a crystal by the application of a voltage.

B.   Mechanical deformation of a crystal by the application of a magnetic field.

C.   The generation of electrical energy by the application of light.

D.   Reversed conduction states when a P-N junction is exposed to light.

3-28C6   In which oscillator circuit would you find a quartz crystal?

A.   Hartley.

B.   Pierce

C.   Colpitts.

D.   All of the above.

3-31D1   What frequency synthesizer circuit uses a phase comparator, look-up table, digital-to-analog converter, and a low-pass antialias filter?

A.   A direct digital synthesizer.

B.   Phase-locked-loop synthesizer.

C.   A diode-switching matrix synthesizer.

D. A hybrid synthesizer.

3-31D2 A circuit that compares the output of a voltage-controlled oscillator (VCO) to a frequency standard and produces an error voltage that is then used to adjust the capacitance of a varactor diode used to control frequency in that same VCO is called what?

A. Doubly balanced mixer.

B. **Phase-locked loop.**

C. Differential voltage amplifier.

D. Variable frequency oscillator.

3-31D4 What spectral impurity components might be generated by a phase-locked-loop synthesizer?

A. Spurs at discrete frequencies.

B. Random spurs which gradually drift up in frequency.

C. **Broadband noise.**

D. Digital conversion noise.

3-31D5 In a direct digital synthesizer, what are the unwanted components on its output?

A. Broadband noise.

B. **Spurs at discrete frequencies.**

C. Digital conversion noise.

D. Nyquist limit noise pulses.

3-31D6 What is the definition of a phase-locked loop (PLL) circuit?

A. A servo loop consisting of a ratio detector, reactance modulator, and voltage-controlled oscillator.

B.   A circuit also known as a monostable multivibrator.

C.   A circuit consisting of a precision push-pull amplifier with a differential input.

D.   A servo loop consisting of a phase detector, a low-pass filter and voltage-controlled oscillator.

3-39E2   What does the term "DAC" refer to in a microprocessor circuit?

A.   Dynamic access controller.

B.   Digital to analog converter.

C.   Digital access counter.

D.   Dial analog control.

3-40E4   What integrated circuit device converts an analog signal to a digital signal?

A.   DAC

B.   DCC

C.   ADC

D.   CDC

3-40E5   What integrated circuit device converts digital signals to analog signals?

A.   ADC

B.   DCC

C.   CDC

D. DAC

3-43F1 Why is the Colpitts oscillator circuit commonly used in a VFO (variable frequency oscillator)?

A. It can be phase locked.

B. It can be remotely tuned.

C. It is stable.

D. It has little or no effect on the crystal's stability.

3-43F2 What is the oscillator stage called in a frequency synthesizer?

A. VCO.　　　　　　　　　　C. Phase detector.

B. Divider.　　　　　　　　　　D. Reference standard.

3-43F3 What are three major oscillator circuits found in radio equipment?

A. Taft, Pierce, and negative feedback.

B. Colpitts, Hartley, and Taft.

C. Taft, Hartley, and Pierce.

D. Colpitts, Hartley, and Pierce.

3-43F4 Which type of oscillator circuit is commonly used in a VFO (variable frequency oscillator)?

A. Colpitts.　　　　　　　　C. Hartley.

B. Pierce.　　　　　　　　　　D. Negative feedback.

3-43F5 What condition must exist for a circuit to oscillate? It must:

A. Have a gain of less than 1.

B. Be neutralized.

C. Have sufficient negative feedback.

D. Have sufficient positive feedback.

18: RECEIVERS

You need to be familiar with a **superheterodyne** receiver.
Heterodyning means the mixing of two signals, and a superheterodyne receiver works by mixing the incoming signal with a signal from the receiver's own **local oscillator (LO)**. The part of the receiver that combines the two signals is called the **mixer**. It combines the incoming signal with the local oscillator and outputs the **intermediate frequency (IF),** which is the difference between the two inputs. For example, in one question on the test, the input signal is 200 MHz, the local oscillator is 150 Mhz. In this example, the IF would be 50 MHz. This question, however, asks you to identify all of the signals that would be produced, prior to any filtering. So the answer includes the IF, 50 MHz, the original signals, 150 MHz and 200 MHz, and also the sum of the two signals, 350 MHz.

When designing a receiver, the selection of the IF depends mostly upon image rejection and selectivity. These two concepts are discussed below.

A block diagram of a superheterodyne receiver is shown below. In this diagram, symbol 2 is the local oscillator, and symbol 1 is the mixer. If you wanted to measure the signal strength of an incoming signal, this could be done with a DC voltmeter in block 4, the AVC.

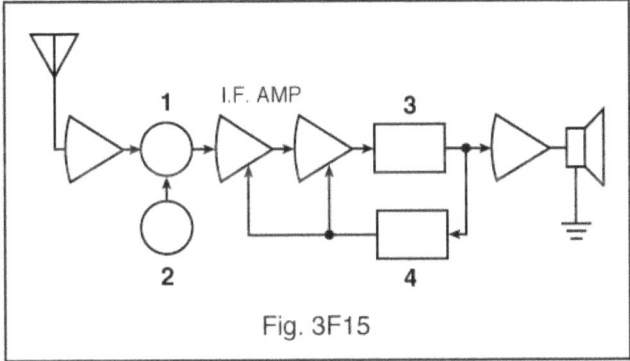

Fig. 3F15

One fact you need to know about the mixer stage is that if excessive amounts of signal are present, the result will be spurious mixer products being generated.

One fact that you need to know about the IF stage is that a receiver SAW IF filter can be used in micro-miniature electronic circuits.

One type of interference that is covered on the test is **image response**. This is caused by the mixer "seeing" an input frequency which is two times the IF lower (or higher) than the desired frequency. For example, on the test, there is a question about listening to a signal on 14.255 MHz on a receiver with an IF of 455 kHz. It is receiving interference from a signal on 13.345 MHz, which is 910 kHz (2 times 455) lower than the desired signal. This is probably an image response. The RF stage of the receiver is normally used to remove the image signal.

The simplest possible superheterodyne receiver would contain an HF oscillator, a mixer, and a detector. The **mixer** is the part that processes signals from the RF amplifier and local oscillator and sends them to the IF filter. In an SSB receiver, the product detector is the part that combines the signals from the IF amplifier and BFO and sends them to the AF amplifier. A **notch filter** is used to reduce interference from a carrier in the receiver passband by "notching it out".

The **discriminator** is the part of an FM receiver that converts signals coming from the IF amplifier to audio.

The **sensitivity** of a receiver is a measure of how weak a signal it can receive. The main limitation on sensitivity is the noise floor of the receiver. The noise figure is the level of noise generated by the receiver itself. The noise figure is primarily determined by the RF stage of the receiver. In other words, the primary purpose of the RF amplifier is to improve the noise figure. When setting the gain of the RF amplifier, it is important to use enough gain so that weak signals overcome the noise generated later in the receiver (primarily by the first mixer). Using too much gain (especially in a VHF-FM receiver) can result in intermodulation interference from nearby transmitters.

Cross-modulation interference is unwanted modulation from a very strong signal being superimposed upon the desired signal.

In VHF receivers, the RF preamplifier is often a GaAsFET type, because this type of amplifier has high gain and a low noise floor.

Sensitivity can be affected by the phenomenon of **desensing**. This means the reduction of sensitivity caused by a strong signal on a nearby frequency.

Another term whose definition you should know for the exam is **dynamic range**. This means the ratio between the strongest tolerable signal and the minimum discernible signal.

Selectivity is a measure of a receiver's ability to separate two signals that are close in frequency. In the front end of a receiver, selectivity can be increased by using a preselector.

The main advantage of a superheterodyne receiver is that only a single frequency, the IF, needs to be amplified. Therefore, the IF amplifier can be optimized for a single frequency. The main purpose of the final IF amplifier stage is to increase the receiver's gain. The purpose of the first IF amplifier is usually to increase selectivity.

A receiver's detector is the part of the receiver that recovers intelligence from the modulated signal. The type of detector used in a Single Sideband (SSB) receiver is a product detector. It uses a mixing process with a locally generated carrier. In an FM receiver, the detector is called the frequency discriminator, or simply **discriminator**.

Some receivers (and transmitters) have a **digital signal processor (DSP)** to remove noise from received signals. The advantage of a DSP filter over an analog filter is that a wide range of filter bandwidths and shapes can be created. A DSP filter can also be set to automatically "notch out" an interfering carrier. DSP filtering is accomplished by converting the signal from analog to digital and using digital processing. ("Digital" is simply another word for "numbers", so don't be confused by one question that uses the word "numbers".) Therefore, a DSP filter contains all of the following: an analog to digital converter, a digital to analog converter, and a digital processor chip. The term **adapting filtering and autocorrelation is the DSP term for noise subtraction.** DSP can be used to control frequency, control modulation, and control detection. DSP is also used in RADAR systems for improved weak signal or target enhancement.

A **software-defined radio (SDR)** is a radio in which most major signal processing functions are performed by software. A software-defined radio will often have a built-in self-test feature, which is usually the best way to make a quick overall evaluation of the radio's condition.

There is one question on the test asking where in a receiver there might be a low-pass filter. The correct answer is that there would be such filters in the power supply and AVC (and probably other places as well), but not in the oscillator.

Finally, you need to know the purpose of de-emphasis in a receiver audio stage. This would be added if a transmitter had pre-emphasis. If you're familiar with the old Dolby noise reduction system for analog tapes, it's the same general concept.

3-31D3 RF input to a mixer is 200 MHz and the local oscillator frequency is 150 MHz. What output would you expect to see at the IF output prior to any filtering?

   A.   50, 150, 200 and 350 MHz.

   B.   50 MHz.

   C.   350 MHz.

   D.   50 and 350 MHz.

3-41F1 What is the limiting condition for sensitivity in a

communications receiver?

A. The noise floor of the receiver.

B. The power supply output ripple.

C. The two-tone intermodulation distortion.

D. The input impedance to the detector.

3-41F2 What is the definition of the term "receiver desensitizing"?

A. A burst of noise when the squelch is set too low.

B. A reduction in receiver sensitivity because of a strong signal on a nearby frequency.

C. A burst of noise when the squelch is set too high.

D. A reduction in receiver sensitivity when the AF gain control is turned down.

3-41F3 What is the term used to refer to a reduction in receiver sensitivity caused by unwanted high-level adjacent channel signals?

A. Desensitizing.

B. Intermodulation distortion.

C. Quieting.

D. Overloading.

3-41F4 What is meant by the term noise figure of a communications receiver?

A. The level of noise entering the receiver from the antenna.

B. The relative strength of a received signal 3 kHz removed from the carrier frequency.

C.   The level of noise generated in the front end and succeeding stages of a receiver.

D.   The ability of a receiver to reject unwanted signals at frequencies close to the desired one.

3-41F5 Which stage of a receiver primarily establishes its noise figure?

A.   The audio stage.

B.   The RF stage.

C.   The IF strip.

D.   The local oscillator.

3-41F6 What is the term for the ratio between the largest tolerable receiver input signal and the minimum discernible signal?

A.   Intermodulation distortion.

B.   Noise floor.

C.   Noise figure.

D.   Dynamic range.

3-42F1 How can selectivity be achieved in the front-end circuitry of a communications receiver?

A.   By using an audio filter.

B.   By using an additional RF amplifier stage.

C.   By using an additional IF amplifier stage.

D.   By using a preselector.

3-42F2 What is the primary purpose of an RF amplifier in a receiver?

A. To provide most of the receiver gain.

B. To vary the receiver image rejection by utilizing the AGC.

C. To improve the receiver's noise figure.

D. To develop the AGC voltage.

3-42F3 How much gain should be used in the RF amplifier stage of a receiver?

A. Sufficient gain to allow weak signals to overcome noise generated in the first mixer stage.

B. As much gain as possible short of self oscillation.

C. Sufficient gain to keep weak signals below the noise of the first mixer stage.

D. It depends on the amplification factor of the first IF stage.

3-42F4 Too much gain in a VHF receiver front end could result in this:

A. Local signals become weaker.

B. Difficult to match receiver impedances.

C. Dramatic increase in receiver current.

D. Susceptibility of intermodulation interference from nearby transmitters.

3-42F5 What is the advantage of a GaAsFET preamplifier in a modern VHF radio receiver?

A. Increased selectivity and flat gain.

B. Low gain but high selectivity.

C. High gain and low noise floor.

D. High gain with high noise floor.

3-42F6 In what stage of a VHF receiver would a low noise amplifier be most advantageous?

A. IF stage.

B. Front end RF stage.

C. Audio stage.

D. Power supply.

3-43F6 In Figure 3F15, which block diagram symbol (labeled 1 through 4) is used to represent a local oscillator?

A. 1   C. 3

B. 2   D. 4

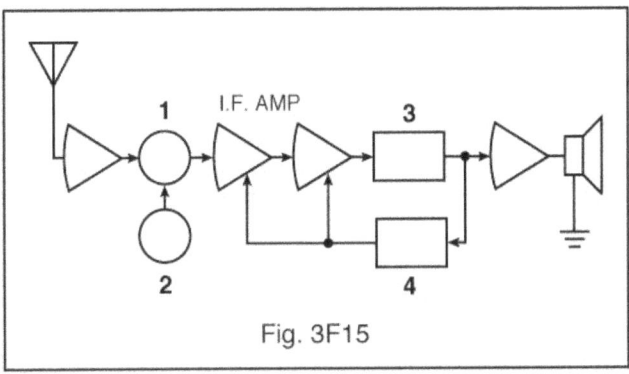

Fig. 3F15

3-44F1 What is the image frequency if the normal channel is 151.000 MHz, the IF is operating at 11.000 MHz, and the LO is at 140.000 MHz?

A. 131.000 MHz.

B. 129.000 MHz.

C. 162.000 MHz.

D.   150.000 MHz.

3-44F2 What is the mixing process in a radio receiver?

A.   The elimination of noise in a wideband receiver by phase comparison.

B.   The elimination of noise in a wideband receiver by phase differentiation.

C.   Distortion caused by auroral propagation.

D.   The combination of two signals to produce sum and difference frequencies.

3-44F3 In what radio stage is the image frequency normally rejected?

A.   RF.

B.   IF.

C.   LO.

D.   Detector.

3-44F4 What are the principal frequencies that appear at the output of a mixer circuit?

A.   Two and four times the original frequency.

B.   The sum, difference and square root of the input frequencies.

C.   The original frequencies and the sum and difference frequencies.

D.   1.414 and 0.707 times the input frequency.

3-44F5 If a receiver mixes a 13.8 MHz VFO with a 14.255 MHz receive signal to produce a 455 kHz intermediate frequency signal, what type of interference will a 13.345 MHz signal produce in the receiver?

A. Local oscillator interference.

B. An image response.

C. Mixer interference.

D. Intermediate frequency interference.

3-44F6 What might occur in a receiver if excessive amounts of signal energy overdrive the mixer circuit?

A. Automatic limiting occurs.

B. Mixer blanking occurs.

C. Spurious mixer products are generated.

D. The mixer circuit becomes unstable and drifts.

3-45F2 Which one of these filters can be used in micro-miniature electronic circuits?

A. High power transmitter cavity.

B. Receiver SAW IF filter.

C. Floppy disk controller.

D. Internet DSL to telephone line filter.

3-46F1 What is the primary purpose of the final IF amplifier stage in a receiver?

A. Dynamic response.

B. Gain.

C. Noise figure performance.

D. Bypass undesired signals.

3-46F2 What factors should be considered when selecting an intermediate frequency?

  A. Cross-modulation distortion and interference.

  B. Interference to other services.

  C. Image rejection and selectivity.

  D. Noise figure and distortion.

3-46F3 What is the primary purpose of the first IF amplifier stage in a receiver?

  A. Noise figure performance.

  B. Tune out cross-modulation distortion.

  C. Dynamic response.

  D. Selectivity.

3-48F1 What is a product detector?

  A. It provides local oscillations for input to the mixer.

  B. It amplifies and narrows the band-pass frequencies.

  C. It uses a mixing process with a locally generated carrier.

  D. It is used to detect cross-modulation products.

3-48F2 Which circuit is used to detect FM-phone signals?

  A. Balanced modulator.

  B. Frequency discriminator.

C. Product detector.

D. Phase splitter.

3-48F4 What is a frequency discriminator in a radio receiver?

A. A circuit for detecting FM signals.

B. A circuit for filtering two closely adjacent signals.

C. An automatic band switching circuit.

D. An FM generator.

3-48F6 What is the definition of detection in a radio receiver?

A. The process of masking out the intelligence on a received carrier to make an S-meter operational.

B. The recovery of intelligence from the modulated RF signal.

C. The modulation of a carrier.

D. The mixing of noise with the received signal.

3-49F1 What is the digital signal processing term for noise subtraction circuitry?

A. Adaptive filtering and autocorrelation.

B. Noise blanking.

C. Noise limiting.

D. Auto squelch noise reduction.

3-49F2 What is the purpose of de-emphasis in the receiver audio stage?

A. When coupled with the transmitter pre-emphasis, flat audio is achieved.

B.   When coupled with the transmitter pre-emphasis, flat audio and noise reduction is received.

C.   No purpose is achieved.

D.   To conserve bandwidth by squelching no-audio periods in the transmission.

3-49F6 What radio circuit samples analog signals, records and processes them as numbers, then converts them back to analog signals?

A.   The pre-emphasis audio stage.

B.   The squelch gate circuit.

C.   The digital signal processing circuit.

D.   The voltage controlled oscillator circuit.

3-50F1 Where would you normally find a low-pass filter in a radio receiver?

A.   In the AVC circuit.

B.   In the Oscillator stage.

C.   In the Power Supply.

D.   A and C, but not B.

3-50F3 What is the term used to refer to the condition where the signals from a very strong station are superimposed on other signals being received?

A.   Intermodulation distortion.

B.   Cross-modulation interference.

C.   Receiver quieting.

D.   Capture effect.

3-50F4 What is cross-modulation interference?

A.   Interference between two transmitters of different modulation type.

B.   Interference caused by audio rectification in the receiver preamp.

C.   Modulation from an unwanted signal heard in addition to the desired signal.

D.   Harmonic distortion of the transmitted signal.

3-50F5 In Figure 3F15 at what point in the circuit (labeled 1 through 4) could a DC voltmeter be used to monitor signal strength?

A.   1   C.   3

B.   2   **D.   4**

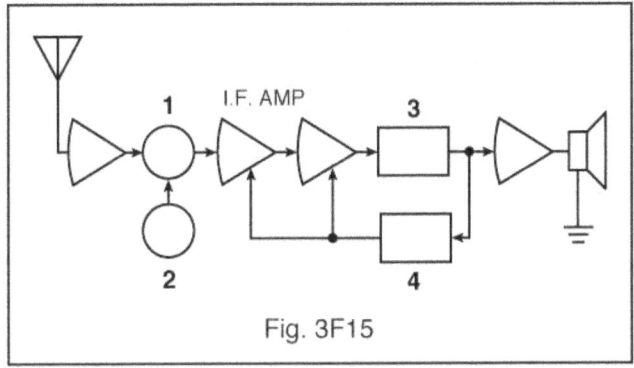

Fig. 3F15

3-81L4 In a software-defined transceiver, what would be the best way for a technician to make a quick overall evaluation of the radio's operational condition?

A.   Set up a spectrum analyzer and service monitor and manually verify the manufacturer's specifications.

B.   Use another radio on the same frequency to check the transmitter.

C.   Use the built-in self-test feature.

D.   Using on-board self-test routines are strictly prohibited by the FCC in commercial transmitters. Amateur Radio is the only service currently authorized to use them.

3-84M1   What is a SDR?

A.   Software Deviation Ratio.

B.   Software Defined Radio.

C.   SWR Meter.

D.   Static Dynamic Ram.

3-84M2   What does the DSP not do in a modern DSP radio?

A.   Control frequency.

B.   Control modulation.

C.   Control detection.

D.   Control SWR.

3-91O5   Digital signal processing (DSP) of RADAR signals (compared with analog) causes:

A.   Improved display graphics.

B.   Improved weak signal or target enhancement.

C.   Less interference with SONAR systems.

D.   Less interference with other radio communications equipment.

19: AMPLIFIERS

The test has a number of questions regarding RF power amplifiers. Most of these questions can be answered very easily be memorizing a few key facts.

Amplifiers are classified as Classes A, AB, B, and C. For the exam, here is all you need to know:

Class A: The output is present for the entire signal cycle (360 degrees). This type of amplifier has the highest linearity (output waveform being identical to the input waveform) and least distortion.

Class AB: The operating angle is more than 180 degrees, but less than 360 degrees. In other words, the output is present for more than half of the signal cycle.

Class B: The output is present for 180 degrees (half) of the input cycle.

Class C: The output is present for less than 180 degrees (less than half) of the signal cycle. In other words, the bias is set well beyond cutoff. This type of amplifier has the greatest efficiency.

As you can see, as you go down the chart, you move from: A greater portion of the input cycle, more linearity, and less efficiency, to: A smaller portion of the input cycle, less linearity, but more efficiency. (One question indirectly asks you about the efficiency of a Class AB amplifier. In this question, the amplifier has an output of 500 watts, and it asks what the input is. You should remember that the Class AB amplifier is inefficient, and pick the answer with the most inefficiency. The correct answer is 1000 watts, meaning that the amplifier is only 50% efficient, which would be typical of a Class AB amplifier.)

For the exam, you need to be familiar with the concept of neutralization. To neutralize an amplifier, you adjust a capacitor near the output. This prevents the tube or transistor from oscillating, which would cause parasitic signals. You should also be aware that a push-pull amplifier reduces even-order harmonics.

3-51G1 What class of amplifier is distinguished by the presence of output throughout the entire signal cycle and the input never goes into the cutoff region?

   A. **Class A.**

   B. Class B.

   C. Class C.

   D. Class D.

3-51G2 What is the distinguishing feature of a Class A amplifier?

   A. Output for less than 180 degrees of the signal cycle.

   B. Output for the entire 360 degrees of the signal cycle.

   C. Output for more than 180 degrees and less than 360 degrees of the signal cycle.

   D. Output for exactly 180 degrees of the input signal cycle.

3-51G3 Which class of amplifier has the highest linearity and least distortion?

   A. **Class A.**

   B.   Class B.

   C.   Class C.

   D.   Class AB.

3-51G4 Which class of amplifier provides the highest efficiency?

   A.   Class A.

   B.   Class B.

   C.   Class C.

   D.   Class AB.

3-51G5 What class of amplifier is distinguished by the bias being set well beyond cutoff?

   A.   Class A.

   B.   Class C.

   C.   Class B.

   D.   Class AB.

3-51G6 Which class of amplifier has an operating angle of more than 180 degrees but less than 360 degrees when driven by a sine wave signal?

   A.   Class A.

   B.   Class B.

   C.   Class C.

   D.   Class AB.

3-52G1 The class B amplifier output is present for what portion of the input cycle?

    A. 360 degrees.

    B. Greater than 180 degrees and less than 360 degrees.

    C. Less than 180 degrees.

    D. 180 degrees.

3-52G3 The class C amplifier output is present for what portion of the input cycle?

    A. Less than 180 degrees.

    B. Exactly 180 degrees.

    C. 360 degrees.

    D. More than 180 but less than 360 degrees.

3-52G4 What is the approximate DC input power to a Class AB RF power amplifier stage in an unmodulated carrier transmitter when the PEP output power is 500 watts?

    A. 250 watts.

    B. 600 watts.

    C. 800 watts.

    D. 1000 watts.

3-52G5 The class AB amplifier output is present for what portion of the input cycle?

    A. Exactly 180 degrees.

B. 360 degrees

C. More than 180 but less than 360 degrees.

D. Less than 180 degrees.

3-52G6 What class of amplifier is characterized by conduction for 180 degrees of the input wave?

A. Class A.

B. Class B.

C. Class C.

D. Class D.

3-56G2 How can parasitic oscillations be eliminated in a power amplifier?

A. By tuning for maximum SWR.

B. By tuning for maximum power output.

C. By neutralization.

D. By tuning the output.

3-56G5 How can even-order harmonics be reduced or prevented in transmitter amplifier design?

A. By using a push-push amplifier.

B. By operating class C.

C. By using a push-pull amplifier.

D. By operating class AB.

20: FILTERS

There are some basic questions on the exam covering filters.

Most filters are made of passive components, such as capacitors and inductors. There are three general groups of filters, and their functions are obvious from their names. A **high-pass filter** allows high frequencies to pass, but blocks low frequencies. A **low-pass filter** allows low frequencies to pass, but blocks high frequencies. A **band-pass** filter allows signals within a particular band to pass, but blocks frequencies that are higher and lower. A **band-stop** filter would be the opposite of a band-pass filter. It would block frequencies of a particular band, but let higher and lower frequencies pass.

There are a few questions about particular filters. You need to know that a Chebyshev filter allows ripple in the passband. You also need to know that you would use an m-derived filter, and not a constant-k filter, if you need more attenuation at a certain frequency, and that frequency is close to the constant-k filter's cutoff frequency. On the other hand, a constant-k filter has high attenuation at distant frequencies. Finally, you need to know that a Butterworth filter has a maximally flat response over its passband.

One question on the test seems too simple, but it really is simple, and not a trick question. If the filter bandwidth of a receiver IF section is too wide, then undesired signals will reach the audio stage.

Active filters use operational amplifiers rather than passive components. The advantage of this type of filter is that it exhibits gain rather than loss. There is one question about operational amplifier filters that seems to simple, but the simple answer is correct. When designing an audio filter using an op amp, the most important parameter is the bandpass characteristics. In other words, if you are designing a filter, you need to figure out what you want to filter out!

3-30D1 What is the main advantage of using an op-amp audio filter over a passive LC audio filter?

    A. Op-amps are largely immune to vibration and temperature change.

    B. Most LC filter manufacturers have retooled to make op-amp filters.

    C. Op-amps are readily available in a wide variety of operational voltages and frequency ranges.

    D. Op-amps exhibit gain rather than insertion loss.

3-45F5 What is an undesirable effect of using too wide a filter bandwidth in the IF section of a receiver?

    A. Output-offset overshoot.

    B. Undesired signals will reach the audio stage.

    C. Thermal-noise distortion.

    D. Filter ringing.

3-46F4 What parameter must be selected when designing an audio filter using an op-amp?

    A. Bandpass characteristics.

    B. Desired current gain.

    C. Temperature coefficient.

D. Output-offset overshoot.

3-46F5 What are the distinguishing features of a Chebyshev filter?

A. It has a maximally flat response over its passband.

B. It only requires inductors.

C. It allows ripple in the passband.

D. A filter whose product of the series- and shunt-element impedances is a constant for all frequencies.

3-46F6 When would it be more desirable to use an m-derived filter over a constant-k filter?

A. When the response must be maximally flat at one frequency.

B. When the number of components must be minimized.

C. When high power levels must be filtered.

D. When you need more attenuation at a certain frequency that is too close to the cut-off frequency for a constant-k filter.

3-47F2 Which statement is true regarding the filter output characteristics shown in Figure 3F16?

A. C is a low pass curve and B is a band pass curve.

B. B is a high pass curve and D is a low pass curve.

C. A is a high pass curve and B is a low pass curve.

D. A is a low pass curve and D is a band stop curve.

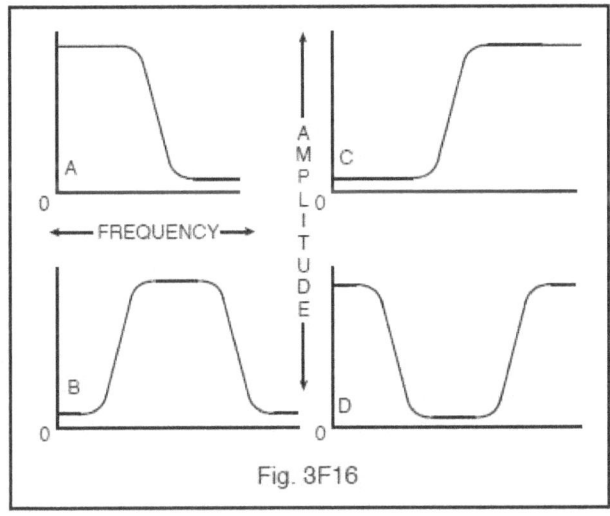

Fig. 3F16

3-47F3 What are the three general groupings of filters?

A. High-pass, low-pass and band-pass.

B. Inductive, capacitive and resistive.

C. Audio, radio and capacitive.

D. Hartley, Colpitts and Pierce.

3-47F4 What is an m-derived filter?

A. A filter whose input impedance varies widely over the design bandwidth.

B. A filter whose product of the series- and shunt-element impedances is a constant for all frequencies.

C. A filter whose schematic shape is the letter "M".

D. A filter that uses a trap to attenuate undesired frequencies too near cutoff for a constant-k filter.

3-47F5 What is an advantage of a constant-k filter?

A. It has high attenuation of signals at frequencies far removed from the pass band.

B. It can match impedances over a wide range of frequencies.

C. It uses elliptic functions.

D. The ratio of the cutoff frequency to the trap frequency can be varied.

3-47F6 What are the distinguishing features of a Butterworth filter?

A. A filter whose product of the series- and shunt-element impedances is a constant for all frequencies.

B. It only requires capacitors.

C. It has a maximally flat response over its passband.

D. It requires only inductors.

21: OPERATIONAL AMPLIFIERS

An operational amplifier (op-amp) is a high-gain voltage amplifier. Its symbol is shown below:

OP AMP

The output is the terminal at the right, and there are two inputs to the left. The output is proportional to the difference between the two voltages at the input: The voltage at the terminal marked + minus the voltage at the terminal marked negative. The terminal marked + is called the non-inverting input. The terminal with the minus sign is called the inverting input.

The gain of an op-amp by itself is extremely high. To keep the gain at usable levels, there is usually some sort of feedback from the output to the input. This external feedback network determines the circuit's gain. Because the purpose of this feedback is to reduce gain, it goes from the output to the inverting input. When an op-amp is used without a feedback network, or open-loop, the circuit is called a comparator, because it is being used merely to compare the two inputs.

In some of the problems on the test, there will be an input voltage at one terminal only, and the other terminal will be grounded. In these cases, you will be able to easily eliminate wrong answers. If the voltage is at the negative terminal, then the output will be the negative of the input. If the input voltage is at the positive terminal, then the output voltage will have the same sign as the input. And since an op-amp is a voltage amplifier, the voltage will increase.

The output of an op amp is always 180 degrees out of phase from the input.

An op-amp is sometimes used as a timing oscillator. One such circuit is shown below. On this diagram, you need to be aware that if the value of the capacitor is increased, the frequency will be lower.

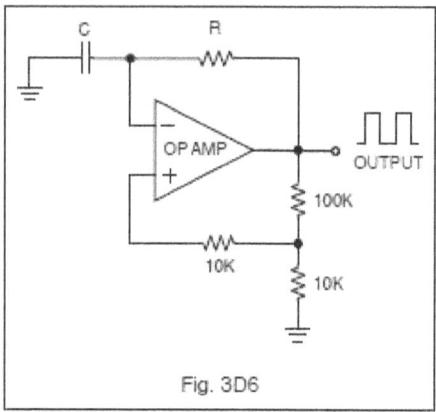

Fig. 3D6

3-30D2 What are the characteristics of an inverting operational amplifier (op-amp) circuit?

    A.   It has input and output signals in phase.

    B.   Input and output signals are 90 degrees out of phase.

    C.   It has input and output signals 180 degrees out of phase.

    D.   Input impedance is low while the output impedance is high.

3-30D3 Gain of a closed-loop op-amp circuit is determined by?

    A. The maximum operating frequency divided by the square root of the load impedance.

    B. The op-amp's external feedback network.

    C. Supply voltage and slew rate.

    D. The op-amp's internal feedback network.

3-30D4 Where is the external feedback network connected to control the gain of a closed-loop op-amp circuit?

    A. Between the differential inputs.

    B. From output to the non-inverting input.

    C. From output to the inverting input.

    D. Between the output and the differential inputs.

3-30D5 Which of the following op-amp circuits is operated open-loop?

    A. Non-inverting amp.   C. Active filter.

    B. Inverting amp.   **D. Comparator.**

3-30D6 In the op-amp oscillator circuit shown in Figure 3D6, what would be the most noticeable effect if the capacitance of C were suddenly doubled?

    A. Frequency would be lower.

    B. Frequency would be higher.

    C. There would be no change. The inputs are reversed, therefore the circuit cannot function.

    D. None of the above.

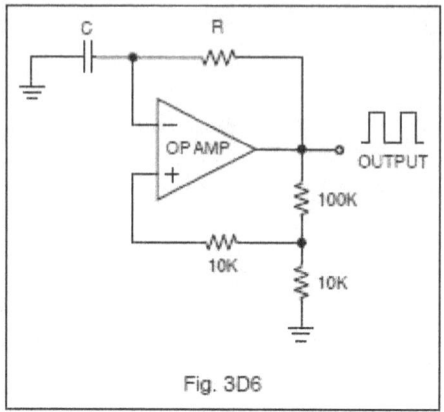

Fig. 3D6

3-32D1 Given the combined DC input voltages, what would the output voltage be in the circuit shown in Figure 3D7?

A. 150 mV

B. 5.5 V

C. -15 mv

D. -5.5 V

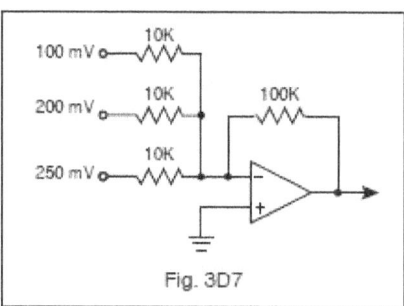

Fig. 3D7

22: SINGLE SIDEBAND AND AM

Because single sideband (**SSB**) is widely used for HF marine and aviation communication, there are a number of questions on the exam regarding this emission type. As the name implies, an SSB signal contains only a single sideband (an AM signal contains two sidebands). Unlike AM, no carrier is transmitted.

A block diagram of an SSB transmitter is shown below:

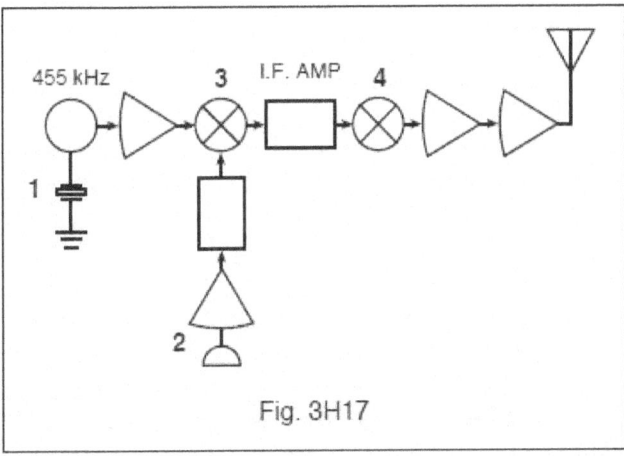

Fig. 3H17

The giveaway that this is an SSB transmitter is the filter, labeled 4 on the diagram. The signal coming out of the I.F. amp is double sideband, and the filter removes the unwanted sideband. The audio is inserted at point

2 on the diagram. The modulator in an SSB transmitter is called a balanced modulator. It is "balanced" because it is creating a double sideband signal, and one of those sidebands is later filtered out. For the test, you need to know the proper terminology. The carrier is **cancelled**, and the unwanted sideband is **filtered** out.

One characteristic of SSB modulation is that the power output varies in proportion to the amount of modulation. In other words, if speech is being transmitted, the power output of the transmitter is zero when the speaker is silent. As the speech gets louder, the power output increases. Once you know this fact, you can easily answer a number of questions: The ration of PEP (peak power) to average power depends on the speech characteristics. The normal average is about 2.5 to 1. You don't need to know the exact number, because the other answers are clearly wrong, as long as you understand that power rises and falls with the voice. If there is power output when there is no speech, this means that there is something wrong with the transmitter which is causing the carrier to be transmitted. Also, because the power varies with the speech, the power supply voltage can fluctuate. Therefore, a low-voltage indicator light might come on with each syllable.

You should know that in the FCC rules, SSB transmissions are referred to as type J3E emission. (If you forget this, you can figure out the correct answer, because the SSB signal is also correctly described in the answer.)

The common test for linearity of the SSB output is the two-tone test. The signal is modulated with two AF sine waves, and the result is observed on the oscilloscope. The two tones must not be harmonically related. In other words, one frequency cannot be a multiple of the other one.

To amplify an SSB signal, a linear amplifier must be used. Linear means that the output waveform is identical to the input. To evaluate the linearity of an amplifier (since it's linear, the question asks about a class A amplifier), the most important piece of information you need to know is the peak voltage of the input signal. If the output signal remains linear on the peaks, then it should be linear elsewhere. If a non-linear amplifier is used for SSB, the result will be distortion.

The one question on the test about standard AM modulation (what the test calls double-sideband phone) refers to the fact that the AM modulation is generally added by modulating the supply voltage of the final amplifier. Since the signal is not modulated up to that point, a non-linear amplifier can be used. Therefore, the correct answer is that AM can be generated by modulating the voltage of a class C (non-linear) final amplifier.

3-52G2 What input-amplitude parameter is most valuable in evaluating the signal-handling capability of a Class A amplifier?

A. Average voltage.

B. RMS voltage.

C. Peak voltage.

D. Resting voltage.

3-53G5 How can a single-sideband phone signal be generated?

A. By driving a product detector with a DSB signal.

B. By using a reactance modulator followed by a mixer.

C. By using a loop modulator followed by a mixer.

D. By using a balanced modulator followed by a filter.

3-53G6 What is a balanced modulator?

A. An FM modulator that produces a balanced deviation.

B. A modulator that produces a double sideband, suppressed carrier signal.

C. A modulator that produces a single sideband, suppressed carrier signal.

D. A modulator that produces a full carrier signal.

3-55G1 What will occur when a non-linear amplifier is used with a single-sideband phone transmitter?

    A. Reduced amplifier efficiency.

    B. Increased intelligibility.

    C. Sideband inversion.

    D. Distortion.

3-55G2 To produce a single-sideband suppressed carrier transmission it is necessary to ____ the carrier and to ____ the unwanted sideband.

    A. Filter, filter.

    B. Cancel, filter.

    C. Filter, cancel.

    D. Cancel, cancel.

3-55G3 In a single-sideband phone signal, what determines the PEP-to-average power ratio?

    A. The frequency of the modulating signal.

    B. The degree of carrier suppression.

    C. The speech characteristics.

    D. The amplifier power.

3-55G4 What is the approximate ratio of peak envelope power to average power during normal voice modulation peak in a single-sideband phone signal?

    A. 2.5 to 1.

    B. 1 to 1.

C.   25 to 1.

D.   100 to 1.

3-58H1 In Figure 3H17, the block labeled 4 would indicate that this schematic is most likely a/an:

A.   Audio amplifier.

B.   Shipboard RADAR.

C.   SSB radio transmitter.

D.   Wireless LAN (local area network) computer.

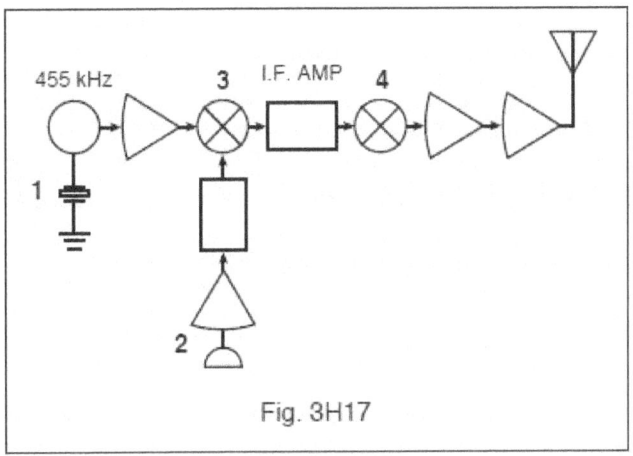

Fig. 3H17

3-58H2 In Figure 3H17, which block diagram symbol (labeled 1 through 4) represents where audio intelligence is inserted?

A.   1   C.   3

B.   2   D.   4

3-58H3 What kind of input signal could be used to test the amplitude linearity of a single-sideband phone transmitter while viewing the output on an oscilloscope?

    A.   Whistling in the microphone.

    B.   An audio frequency sine wave.

    C.   A two-tone audio-frequency sine wave.

    D.   An audio frequency square wave.

3-58H4 What does a two-tone test illustrate on an oscilloscope?

    A.   Linearity of a SSB transmitter.

    B.   Frequency of the carrier phase shift.

    C.   Percentage of frequency modulation.

    D.   Sideband suppression.

3-58H5 How can a double-sideband phone signal be produced?

    A.   By using a reactance modulator.

    B.   By varying the voltage to the varactor in an oscillator circuit.

    C.   By using a phase detector, oscillator, and filter in a feedback loop.

    D.   By modulating the supply voltage to a class C amplifier.

3-58H6 What type of signals are used to conduct an SSB two-tone test?

    A.   Two audio signals of the same frequency, but shifted 90 degrees in phase.

    B.   Two non-harmonically related audio signals that are within the modulation band pass of the transmitter.

    C.   Two different audio frequency square wave signals of equal amplitude.

D. Any two audio frequencies as long as they are harmonically related.

3-76L6 What instrument is used to check the signal quality of a single-sideband radio transmission?

A. Field strength meter.

B. Signal level meter.

C. Sidetone monitor.

D. Oscilloscope.

3-81L1 On a 150 watt marine SSB HF transceiver, what would be indicated by a steady output of 75 watts when keying the transmitter on?

A. There is probably a defect in the system causing the carrier to be transmitted.

B. One of the sidebands is missing.

C. Both sidebands are being transmitting.

D. The operation is normal.

3-82M5 Which of the following statements about SSB voice transmissions is correct?

A. They use A3E emission which are produced by modulating the final amplifier.

B. They use F3E emission which is produced by phase shifting the carrier.

C. They normally use J3E emissions, which consists of one sideband and a suppressed carrier.

D. They may use A1A emission to suppress the carrier.

3-86N1 What is a common occurrence when voice-testing an SSB aboard a boat?

A. Ammeter fluctuates down with each spoken word.

B. Voltage panel indicator lamps may glow with each syllable.

C. Automatic tuner cycles on each syllable.

D. Minimal voltage drop seen at power source.

23: FM

Frequency Modulation (FM) is widely used in the land mobile service and in VHF marine radio. Therefore, it is reasonable that there are a large number of questions covering FM. You will be able to answer most of them, however, merely by understanding some definitions, and by making some relatively easy calculations.

FM can be produced by feeding the audio directly into the oscillator. It can also be produced by means of **phase modulation**. The important fact to know about phase modulation is that the modulation index does not change, even if the carrier frequency changes.

The main terms you need to be familiar with are deviation, modulation index, deviation ratio,

Deviation is the maximum difference in carrier frequency caused by the modulation. You should know that for normal wideband use on VHF and FM, the maximum deviation is normally 5.0 kHz.

The **deviation ratio** is defined as the maximum deviation from the carrier frequency divided by the maximum audio modulating frequency. For example, if the maximum deviation is plus or minus 5 kHz, and the maximum modulation frequency is 3 kHz, then the deviation ratio is 1.66.

The **modulation index** is defined as the ratio between the deviation of the transmitted signal and the modulating audio signal. For example, if the maximum deviation is 3000 Hz and the modulating frequency is 1000 Hz, then the modulation index is 3. (In fact, on both modulation index questions on the exam, the correct answer is 3.)

There are a number of questions on various squelching systems. The squelch on a radio receiver turns off the speaker when no desired signal is being received. There are various types of squelch circuits. Most FM receivers have a squelch control which functions as a result of noise. When noise is present, no signal is being received, and the circuit shuts off the speaker. This is normally called "carrier squelch". On the test, however, it is simply called squelch.

There are two other types of squelch on the exam. The first of these is Digital Coded Squelch. There is only one question on the test, and it asks what makes digital coded squelch work. The correct answer is digital codes. Yes, they really put a question that easy on the test!

The other type of squelch is CTCSS, which stands for continuous tone-coded squelch system. As the name implies, this type of squelch relies upon tones which are continuously transmitted along with the transmitted audio. You might be familiar with this type of system, which is often referred to as "PL". PL, however, is a registered trademark of Motorola for its CTCSS system, and the generic name is instead properly used.

The CTCSS tones, as the name implies, are continuously transmitted along with the rest of the audio. Most of these tones are in the range of human hearing, and need to be filtered out in the receiver. This filtering takes place after the discriminator but before the audio section. If you remember that they are audio tones, the wrong answers to this question will be obvious, since they are all referring to RF portions of the receiver.

The most common instrument used by technicians to test FM transceivers is the **service monitor**. This instrument can monitor frequency and modulation, check sensitivity and distortion, and generate audio tones. Many portable and mobile FM transceivers are programmable, and a laptop computer is often used to program them.

3-48F5 In a CTCSS controlled FM receiver, the CTCSS tone is filtered out after the:

A.   IF stage but before the mixer.

B.   Mixer but before the IF.

C.   IF but before the discriminator.

D.   Discriminator but before the audio section.

3-49F3   What makes a Digital Coded Squelch work?

A.   Noise.

B.   Tones.

C.   Absence of noise.

D.   Digital codes.

3-49F4   What causes a squelch circuit to function?

A.   Presence of noise.

B.   Absence of noise.

C.   Received tones.

D.   Received digital codes.

3-49F5   What makes a CTCSS squelch work?

A.   Noise.

B.   Tones.

C.   Absence of noise.

D.   Digital codes.

3-53G1   What is the modulation index in an FM phone signal having a maximum frequency deviation of 3,000 Hz on either side of the carrier frequency when the modulating frequency is 1,000 Hz?

A. 0.3

B. 3,000

C. 3

D. 1,000

3-53G2 What is the modulation index of a FM phone transmitter producing a maximum carrier deviation of 6 kHz when modulated with a 2 kHz modulating frequency?

A. 3

B. 6,000

C. 2,000

D. 1

3-53G4 How does the modulation index of a phase-modulated emission vary with RF carrier frequency?

A. It does not depend on the RF carrier frequency.

B. Modulation index increases as the RF carrier frequency increases.

C. It varies with the square root of the RF carrier frequency.

D. It decreases as the RF carrier frequency increases.

3-57H1 The deviation ratio is the:

A. Audio modulating frequency to the center carrier frequency.

B. Maximum carrier frequency deviation to the highest audio modulating frequency.

C. Carrier center frequency to the audio modulating frequency.

D. Highest audio modulating frequency to the average audio modulating frequency.

3-57H2 What is the deviation ratio for an FM phone signal having a maximum frequency deviation of plus or minus 5 kHz and accepting a maximum modulation rate of 3 kHz?

A. 60

B. 0.16

C. 0.6

D. 1.66

3-57H3 What is the deviation ratio of an FM-phone signal having a maximum frequency swing of plus or minus 7.5 kHz and accepting a maximum modulation rate of 3.5 kHz?

A. 2.14

B. 0.214

C. 0.47

D. 47

3-57H4 How can an FM-phone signal be produced in a transmitter?

A. By modulating the supply voltage to a class-B amplifier.

B. By modulating the supply voltage to a class-C amplifier.

C. By using a balanced modulator.

D. By feeding the audio directly to the oscillator.

3-57H5 What is meant by the term modulation index?

A. **The ratio between the deviation of a frequency modulated signal and the modulating frequency.**

B. The processor index.

C. The FM signal-to-noise ratio.

D. The ratio of the maximum carrier frequency deviation to the highest audio modulating frequency.

3-57H6 In an FM-phone signal, what is the term for the maximum deviation from the carrier frequency divided by the maximum audio modulating frequency?

A. Deviation index.

B. Modulation index.

C. **Deviation ratio.**

D. Modulation ratio.

3-77L6 What instrument is commonly used by radio service technicians to monitor frequency, modulation, check receiver sensitivity, distortion, and to generate audio tones?

A. Oscilloscope.

B. Spectrum analyzer.

C. **Service monitor.**

D. DMM.

3-78L6 What is the maximum FM deviation for voice operation of a normal wideband channel on VHF and UHF?

A. 2.5 kHz

B. **5.0 kHz**

C.   7.5 kHz

D.   10 kHz

3-81L3 A common method of programming portable or mobile radios is to use a:

A.   A laptop computer.

B   Dummy load.

C.   A wattmeter.

D.   A signal generator.

24: BANDWIDTH

You need to be familiar with the bandwidth used by various radio
signals, and the types of filters that you would use in connection with
those different modes. In general, a receiver's IF filter should have a
bandwidth that is slightly greater than the bandwidth of the signal being
received.

If you know the following facts, you should be able to answer these
questions easily.

A single-sideband phone signal typically has a bandwidth of about 2
kHz. Therefore a 2.1 kHz or 2.4 kHz filter would be a good choice. For
weather fax reception, a somewhat narrower filter would be suitable.
The correct answer on the exam calls for a 1 kHz filter.

Double-sideband AM has a bandwidth of about 10 kHz, so a selectivity
of 10 kHz would be a good choice. (For this question, just remember
that the AM broadcast band in the U.S. is divided up into 10 kHz
channels.)

FM voice signals take up a greater bandwidth, so 15 kHz selectivity is
desirable for FM. There is one question requiring you to compute FM
bandwidth. No calculation is necessary, since the correct answer is the
bandwidth of a typical FM signal, 16 kHz.

To determine the effective bandwidth of a receiver, the test to perform would be a modulation-acceptance bandwidth test.

3-45F1 What degree of selectivity is desirable in the IF circuitry of a wideband FM phone receiver?

 A. 1 kHz.

 B. 2.4 kHz.

 C. 4.2 kHz.

 D. 15 kHz.

3-45F3 A receiver selectivity of 2.4 kHz in the IF circuitry is optimum for what type of signals?

 A. CW.

 B. Double-sideband AM voice.

 C. SSB voice.

 D. FSK RTTY.

3-45F4 A receiver selectivity of 10 KHz in the IF circuitry is optimum for what type of signals?

 A. Double-sideband AM.

 B. SSB voice.

 C. CW.

 D. FSK RTTY.

3-45F6 How should the filter bandwidth of a receiver IF section compare with the bandwidth of a received signal?

 A. Slightly greater than the received-signal bandwidth.

B. Approximately half the received-signal bandwidth.

C. Approximately two times the received-signal bandwidth.

D. Approximately four times the received-signal bandwidth.

3-47F1 A good crystal band-pass filter for a single-sideband phone would be?

A. 5 KHz.   C. 500 Hz.

B. 2.1 KHz.   D. 15 KHz.

3-53G3 What is the total bandwidth of a FM phone transmission having a 5 kHz deviation and a 3 kHz modulating frequency?

A. 3 kHz.

B. 8 kHz.

C. 5 kHz.

D. 16 kHz.

3-78L5 A communications technician would perform a modulation-acceptance bandwidth test in order to:

A. Ascertain the audio frequency response of the receiver.

B. Determine whether the CTCSS in the receiver is operating correctly.

C. Verify the results from a 12 dB SINAD test.

D. Determine the effective bandwidth of a communications receiver.

3-88N3 What would be the bandwidth of a good crystal lattice band-pass filter for weather facsimile HF (high frequency) reception?

A.   500 Hz at -6 dB.

B.   6 kHz at -6 dB.

C.   1 kHz at -6 dB.

D.   15 kHz at -6 dB.

3-88N3 What would be the bandwidth of a good crystal lattice band-pass filter for weather facsimile HF (high frequency) reception?

A.   500 Hz at -6 dB.

B.   6 kHz at -6 dB.

C.   1 kHz at -6 dB.

D.   15 kHz at -6 dB.

25: ANTENNAS AND FEEDLINES

There are relatively few antenna questions on the test, so we won't cover the subject in any particular depth. Knowing a few key concepts will allow you to answer these questions easily. One of the most basic antennas is the half-wave dipole. As the name implies, it is a half wavelength long, and it is fed in the center. For the exam, you need to know that the ends of the antenna are the area with the maximum voltage and minimum current.

The other basic antenna is the quarter-wave vertical. Again, as the name implies, the length of this antenna is approximately one quarter of the wavelength. This type of antenna receives and transmits signals equally from all horizontal directions.

In general, to lengthen an antenna (in other words, make it work on a longer wavelength/lower frequency), you would add inductance in series with the antenna. This stands to reason. An inductor is nothing more than a coil of wire, so you are essentially adding more wire and making it longer. (Loading coils generally reduce the bandwidth of an antenna. Bandwidth is simply the frequency range over which an antenna performs well.) To make an antenna electrically shorter (in other words, to make it work on a shorter wavelength/higher frequency), you would add a capacitor in series with the antenna.

Multiband antennas are often constructed with traps. The trap prevents signals of one frequency from reaching parts of the antenna.

When an antenna is excited (in other words, when it's hooked up to an operating transmitter), it produces both electromagnetic and electrostatic fields. It's important to know an antenna's radiation resistance, because this makes it possible to match it with the transmitter's impedance. Radiation resistance is defined as "equivalent resistance that would dissipate the same amount of power as that radiated from an antenna."

When receiving, the antenna converts the radio waves into an RF voltage, which is sent through the feedline to the receiver. Obviously, this voltage is proportional to the field strength of the radio waves.

Standing wave ratio (SWR) is a measurement of how well an antenna is matched to a transmitter. A perfect SWR is 1:1. The SWR is actually an indirect measure of how much power is being reflected back to the transmitter. The question regarding reflected power is very simple. It states that a 30 watt transmitter has 5 watts being reflected back. In this situation, 25 watts is actually being radiated. Only one possible cause of a high SWR is mentioned on the test, and that is a detuned antenna coupler.

One phenomenon that is relevant to antennas is skin effect. The is the phenomenon where RF current flows in a thin layer close to the surface. The layer gets thinner as the frequency increases. This is why, for example, antennas can be constructed from hollow tubing.

The questions regarding feedlines are very straightforward. You need to know that the velocity factor of a feedline is the ratio of the velocity through that feedline, compared to the speed of light. The velocity factor is affected by dielectrics in the line.

Nitrogen is sometimes used in feedlines to prevent moisture from entering.

A time domain reflectometer (TDR) is an instrument used to measure the characteristics of a transmission line. It consists of an oscilloscope combined with a pulse generator. A frequency domain reflectometer, on the other hand, is used for checking antennas and transmission lines at the particular operating frequency of the antenna.

Finally, an **isolator** is simply a device that allows RF energy to pass in one direction, but absorbs RF going the other direction.

3-3A4 Skin effect is the phenomenon where:

A. RF current flows in a thin layer of the conductor, closer to the surface, as frequency increases.

B. RF current flows in a thin layer of the conductor, closer to the surface, as frequency decreases.

C. Thermal effects on the surface of the conductor increase the impedance.

D. Thermal effects on the surface of the conductor decrease the impedance.

3-63J1 Which of the following could cause a high standing wave ratio on a transmission line?

A. Excessive modulation.

B. An increase in output power.

C. A detuned antenna coupler.

D. Low power from the transmitter.

3-63J2 Why is the value of the radiation resistance of an antenna important?

A. Knowing the radiation resistance makes it possible to match impedances for maximum power transfer.

B. Knowing the radiation resistance makes it possible to measure the near-field radiation density from transmitting antenna.

C. The value of the radiation resistance represents the front-to-side ratio of the antenna.

D. The value of the radiation resistance represents the front-to-back ratio of the antenna.

3-63J3 A radio frequency device that allows RF energy to pass through in one direction with very little loss but absorbs RF power in the opposite direction is a:

    A. Circulator.

    B. Wave trap.

    C. Multiplexer.

    D. Isolator.

3-63J4 What is an advantage of using a trap antenna?

    A. It may be used for multiband operation.

    B. It has high directivity in the high-frequency bands.

    C. It has high gain.

    D. It minimizes harmonic radiation.

3-63J5 What is meant by the term radiation resistance of an antenna?

    A. Losses in the antenna elements and feed line.

    B. The specific impedance of the antenna.

    C. The resistance in the trap coils to received signals.

    D. An equivalent resistance that would dissipate the same amount of power as that radiated from an antenna.

3-63J6 What is meant by the term antenna bandwidth?

    A. Antenna length divided by the number of elements.

    B. The frequency range over which an antenna can be expected to perform well.

C.   The angle between the half-power radiation points.

D.   The angle formed between two imaginary lines drawn through the ends of the elements.

3-64J2 The voltage produced in a receiving antenna is:

A.   Out of phase with the current if connected properly.

B.   Out of phase with the current if cut to 1/3 wavelength.

C.   Variable depending on the station's SWR.

D.   Always proportional to the received field strength.

3-64J3 Which of the following represents the best standing wave ratio (SWR)?

A.   1:1.

B.   1:1.5.

C.   1:3.

D.   1:4.

3-64J4 At the ends of a half-wave antenna, what values of current and voltage exist compared to the remainder of the antenna?

A.   Equal voltage and current.

B.   Minimum voltage and maximum current.

C.   Maximum voltage and minimum current.

D.   Minimum voltage and minimum current.

3-65J1 A vertical 1/4 wave antenna receives signals:

A.   In the microwave band.

B. In one vertical direction.

C. In one horizontal direction.

D. Equally from all horizontal directions.

3-65J2 The resonant frequency of a Hertz antenna can be lowered by:

A. Lowering the frequency of the transmitter.

B. Placing an inductance in series with the antenna.

C. Placing a condenser in series with the antenna.

D. Placing a resistor in series with the antenna.

3-65J3 An excited 1/2 wavelength antenna produces:

A. Residual fields.

B. An electro-magnetic field only.

C. Both electro-magnetic and electro-static fields.

D. An electro-flux field sometimes.

3-65J4 To increase the resonant frequency of a 1/4 wavelength antenna:

A. Add a capacitor in series.

B. Lower capacitor value.

C. Cut antenna.

D. Add an inductor.

3-65J5 What happens to the bandwidth of an antenna as it is shortened through the use of loading coils?

A. It is increased.

B. It is decreased.

C. No change occurs.

D. It becomes flat.

3-65J6 To lengthen an antenna electrically, add a:

A. Coil.

B. Resistor.

C. Battery.

D. Conduit.

3-66J1 What is the meaning of the term velocity factor of a transmission line?

A. The ratio of the characteristic impedance of the line to the terminating impedance.

B. The velocity of the wave on the transmission line divided by the velocity of light in a vacuum.

C. The velocity of the wave on the transmission line multiplied by the velocity of light in a vacuum.

D. The index of shielding for coaxial cable.

3-66J2 What determines the velocity factor in a transmission line?

A. The termination impedance.

B. The line length.

C. Dielectrics in the line.

D. The center conductor resistivity.

3-66J3 Nitrogen is placed in transmission lines to:

A. Improve the "skin-effect" of microwaves.

B. Reduce arcing in the line.

C. Reduce the standing wave ratio of the line.

D. Prevent moisture from entering the line.

3-66J4 A perfect (no loss) coaxial cable has 7 dB of reflected power when the input is 5 watts. What is the output of the transmission line?

A. 1 watt.

B. 1.25 watts.

C. 2.5 watts.

D. 5 watts.

3-74L6 What instrument may be used to verify proper radio antenna functioning?

A. Digital ohm meter.

B. Hewlett-Packard frequency meter.

C. An SWR meter.

D. Different radio.

3-77L1 A(n) _____ and _____ can be combined to measure the characteristics of transmission lines. Such an arrangement is known as a time-domain reflectometer (TDR).

A. Frequency spectrum analyzer, RF generator.

B. Oscilloscope, pulse generator.

C. AC millivolt meter, AF generator.

D. Frequency counter, linear detector.

3-77L4 What instrument is most accurate when checking antennas and transmission lines at the operating frequency of the antenna?

A. Time domain reflectometer.

B. Wattmeter.

C. DMM.

D. Frequency domain reflectometer.

26: INTERFERENCE

There are a few additional questions regarding interference that are not covered in other chapters.

Pulse-type interference which is related to the speed of an engine is probably caused by the engine's ignition system and can be reduced by installing resistances in series with the spark plug wires. Ferrite beads can also be used in the primary and secondary ignition leads.

One good tool for locating interference (referred to in one question as EMI, electromagnetic interference) is a portable AM radio. This is very useful because it is susceptible to interference, and the antenna is usually very directional.

In the chapter on receivers, we discussed the concept of intermodulation interference. This is most commonly experienced on FM, and is the result of two transmitters being in close proximity. The two signals can mix together in one of their final amplifiers, resulting in undesired signals being transmitted on their sum and difference frequencies. This interference can be reduced or eliminated by installing a terminated circulator or ferrite insulator in the feedline.

A shorted stub line can be used to absorb the even harmonics of a signal. This type of stub should be ¼ wavelength long.

3-50F2 How can ferrite beads be used to suppress ignition noise? Install them:

    A.   In the resistive high voltage cable every 2 years.

    B.   Between the starter solenoid and the starter motor.

    C.   Install them in the primary and secondary ignition leads.

    D.   In the antenna lead.

3-50F6 Pulse type interference to automobile radio receivers that appears related to the speed of the engine can often be reduced by:

    A.   Installing resistances in series with spark plug wires.

    B.   Using heavy conductors between the starting battery and the starting motor.

    C.   Connecting resistances in series with the battery.

    D.   Grounding the negative side of the battery.

3-56G1 How can intermodulation interference between two transmitters in close proximity often be reduced or eliminated?

    A.   By using a Class C final amplifier with high driving power.

    B.   By installing a terminated circulator or ferrite isolator in the feed line to the transmitter and duplexer.

    C.   By installing a band-pass filter in the antenna feed line.

    D.   By installing a low-pass filter in the antenna feed line.

3-56G3 What is the name of the condition that occurs when the signals of two transmitters in close proximity mix together in one or both of their final amplifiers, and unwanted signals at the sum and difference frequencies of the original transmissions are generated?

    A.   Amplifier desensitization.

B.   Neutralization.

C.   Adjacent channel interference.

D.   Intermodulation interference.

3-66J5 Referred to the fundamental frequency, a shorted stub line attached to the transmission line to absorb even harmonics could have a wavelength of:

A.   1.41 wavelength.

B.   1/2 wavelength.

C.   1/4 wavelength.

D.   1/6 wavelength.

3-75L3 What equipment may be useful to track down EMI aboard a ship or aircraft?

A.   Fluke multimeter.

B.   An oscilloscope.

C.   Portable AM receiver.

D.   A logic probe.

27: BATTERIES AND POWER SUPPLIES

There are a number of questions on the exam relating to batteries and power supplies. One question shows you the following diagram of a switching power supply and asks you what is wrong with it.

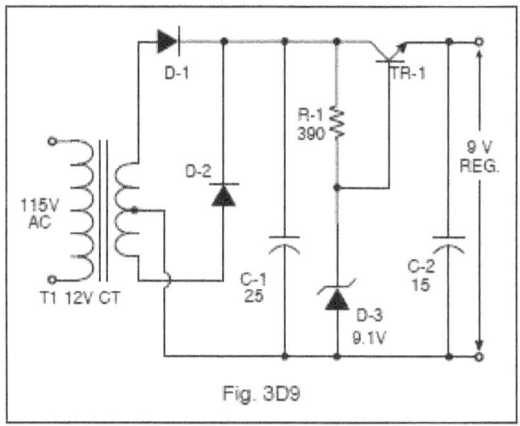

Fig. 3D9

There is nothing wrong with it, and the power supply is perfectly functional. There is one question asking the ratio between the output and input frequencies of a full-wave rectifier. The answer is 2:1. A full-wave rectifier converts the negative portion of the sine wave into a positive voltage. Therefore, if the input is a 60 Hz sine wave, the output is 120 positive pulses per second.

There are a number of battery questions, and a few of these are simple Ohm's Law questions. One question asks you to trickle charge a 12.5 volt battery at 0.5 amps. You have 110 volts DC available, and you need to select a resistor to put in series with the battery. In other words, what you need to do is "get rid of" 97.5 volts by running the current through a resistor. To calculate the value of the resistor, simply use Ohm's law:

R = E / I = 97.5 / 0.5 = 195 ohms.

The other question is only slightly more complicated. It tells you you have a 6 volt battery with an internal resistance of 0.01 ohms hooked up to a 6-volt 3-watt lamp. To do this problem, we first need to calculate the resistance of the bulb. Since it is being sold as a "3 watt" bulb, this means that its normal current rating is 0.5 amps (6 volts x 0.5 Amps = 3 watts). With this information, we can calculate the resistance of the bulb:

R = E / I = 6 / 0.5 = 12 ohms.

The current from the battery needs to pass through this resistance, and also through the internal resistance of the battery. Therefore, the total resistance of the circuit is 12.01 ohms. We can then calculate the circuit current:

I = E / R = 6 / 12.01 = 0.4996 amps. (The correct answer on the test is 0.4995, which appears to be a rounding error.)

There are a number of questions regarding battery capacity, and all of these problems can be solved with simple arithmetic. A battery is rated in amp-hours. A one amp-hour battery can be used at one amp for one hour. A two amp-hour battery can be used at two amps for one hour, or one amp for two hours. (In the real world, it makes a difference how fast you discharge a battery, but for the exam, this is ignored. Also, for purposes of the exam, you can use the battery's total capacity, even though this might be a poor practice in the real world.) In these questions, you'll be given a battery and a load, and need to compute how long the battery will last.

The following question is probably the most difficult, so we'll use it as an example:

What capacity in amperes does a storage battery need to be in order to operate a 50 watt transmitter for 6 hours? Assume a continuous transmitter load of 70% of the key-locked demand of 40 A, and an emergency light load of 1.5 A.

We start out by calculating the transmitter's current needs. With the key down, the transmitter draws 40 amps. According to the question, though, its average current draw is 70% of this value, or 28 amps. In addition, the lights are drawing 1.5 amps, so we have a total of 29.5 amps. So the total number of amp-hours we need is 29.5 x 6 = 177 amp-hours.

In some of these questions, the current needs are not specified. For example, one problem refers to a transmitter that uses 325 watts of power from a 12.6 volt battery. Therefore, before doing this problem, you need to calculate the current, I = P / E = 325 / 12.6 = 25.79 amps.

There are questions regarding percentage of regulation of a regulated power supply. The percentage of regulation is defined as the difference between the full load voltage and the no load voltage, divided by the full load voltage. Just remember that when you do the division, you divide by the **lower** voltage that is given. The two questions on the test are:

No load voltage = 12; load voltage = 10; regulation = (12-10) / 10 = 20%

No load voltage = 250; load voltage = 200; regulation = (250-200) / 200 = 25%

TTL (transistor-transistor logic) is discussed in more detail in another chapter. However, there are a number of very easy questions regarding TTL voltages. As you probably know, the standard for TTL is 5 volts for the "high" state (in other words, a binary "1" or "true", and 0 volts for the "low" state (in other words, a binary "0" or "false".) As long as you know this, you don't need to know the actual tolerance to get the questions right. There are questions asking what the acceptable ranges are. For "low", the acceptable voltage is zero to 0.8 volts. For "high", the acceptable range is 2.0 to 5.5 volts. However, you don't need to know the ranges to get these questions right, because the wrong answers are so far out of range.

3-18B3 Given a power supply with a no load voltage of 12 volts and a full load voltage of 10 volts, what is the percentage of voltage regulation?

A.   17 %

B.   80 %

C.   20 %

D.   83 %

3-18B5 Given a power supply with a full load voltage of 200 volts and a regulation of 25%, what is the no load voltage?

A.   150 volts.

B.   160 volts.

C.   240 volts.

D.   250 volts.

3-21C4 What is the ratio of the output frequency to the input frequency of a single-phase full-wave rectifier?

A.   1:1.

B.   1:2.

C.   2:1.

D.   None of these.

3-32D4 In Figure 3D9, determine if there is a problem with this regulated power supply and identify the problem.

A.   R1 value is too low which would cause excessive base current and instantly destroy TR 1.

B.   D1 and D2 are reversed. The power supply simply would not function.

C.   TR1 is shown as an NPN and must be changed to a PNP.

D. There is no problem with the circuit.

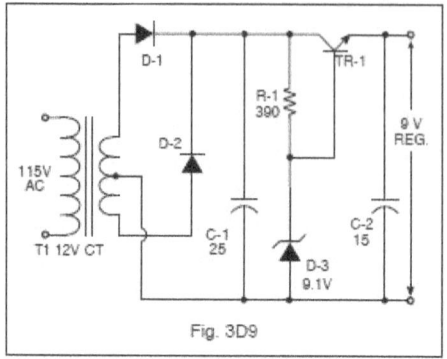

Fig. 3D9

3-33E1 What is the voltage range considered to be valid logic low input in a TTL device operating at 5 volts?

A. 2.0 to 5.5 volts.

B. -2.0 to -5.5 volts.

C. Zero to 0.8 volts.

D. 5.2 to 34.8 volts.

3-33E2 What is the voltage range considered to be a valid logic high input in a TTL device operating at 5.0 volts?

A. 2.0 to 5.5 volts.

B. 1.5 to 3.0 volts.

C. 1.0 to 1.5 volts.

D. 5.2 to 34.8 volts.

3-33E3 What is the common power supply voltage for TTL series integrated circuits?

A. 12 volts.

B.   13.6 volts.

C.   1 volt.

D.   5 volts.

3-60I2 A battery with a terminal voltage of 12.5 volts is to be trickle-charged at a 0.5 A rate. What resistance should be connected in series with the battery to charge it from a 110-V DC line?

A.   95 ohms.

B.   300 ohms.

C.   195 ohms.

D.   None of these.

3-60I3 What capacity in amperes does a storage battery need to be in order to operate a 50 watt transmitter for 6 hours? Assume a continuous transmitter load of 70% of the key-locked demand of 40 A, and an emergency light load of 1.5 A.

A.   100 ampere-hours.

B.   177 ampere-hours.

C.   249 ampere-hours.

D.   None of these.

3-61I1 When an emergency transmitter uses 325 watts and a receiver uses 50 watts, how many hours can a 12.6 volt, 55 ampere-hour battery supply full power to both units?

A.   1.8 hours.

B.   6 hours.

C.   3 hours.

D. 1.2 hours.

3-6I3 A ship RADAR unit uses 315 watts and a radio uses 50 watts. If the equipment is connected to a 50 ampere-hour battery rated at 12.6 volts, how long will the battery last?

A. 1 hour 43 minutes.

B. 28.97 hours.

C. 29 minutes.

D. 10 hours, 50 minutes.

3-6I4 If a marine radiotelephone receiver uses 75 watts of power and a transmitter uses 325 watts, how long can they both operate before discharging a 50 ampere-hour 12 volt battery?

A. 40 minutes.

B. 1 hour.

C. 1 1/2 hours.

D. 6 hours.

3-6I6 A 12.6 volt, 8 ampere-hour battery is supplying power to a receiver that uses 50 watts and a RADAR system that uses 300 watts. How long will the battery last?

A. 100.8 hours.

B. 27.7 hours.

C. 1 hour.

D. 17 minutes or 0.3 hours.

3-6I2 What current will flow in a 6 volt storage battery with an internal resistance of 0.01 ohms, when a 3-watt, 6-volt lamp is connected?

A. 0.4885 amps.

B. 0.4995 amps.

C. 0.5566 amps.

D. 0.5795 amps.

3-6I15 A 6 volt battery with 1.2 ohms internal resistance is connected across two light bulbs in parallel whose resistance is 12 ohms each. What is the current flow?

A. 0.57 amps.

B. 0.83 amps.

C. 1.0 amps.

D. 6.0 amps.

28: MOTORS AND GENERATORS

The exam contains a few questions relating to motors and generators. Even if you have no familiarity with motors and generators, you can easily answer these questions by memorizing a few facts and using a bit of logic.

First of all, you need to know that one horsepower is the same as 746 watts. This question might appear on the test directly. But even if it doesn't, you'll need to know it to perform some calculations.

You should also know the following facts:

1. If the load is removed from an operating DC series motor, this will cause the motor to speed up until it falls apart.
2. If a shunt motor has its shunt field opened, this will cause the motor to speed up.
3. In the context of shunt motors, "voltage regulation" is defined as voltage fluctuations from load to no-load.
4. The output of an AC generator running at a constant speed can be adjusted by varying the amount of field current.

There are only two questions requiring computations. If you're not familiar with motors, you might decide to simply memorize the answers. But by applying a bit of logic, you can easily solve them. We'll work through both of them.

> What is the line current of a 7 horsepower motor operating on 120 volts at full load, a power factor of 0.8, and 95% efficient?

First of all, this is a 7 horsepower motor. One horsepower equals 746 watts, so the amount of power being delivered by the motor is 7 x 742 = 5222 watts. The motor is 95% efficient. This means that the electric power going into the motor is more than the mechanical power coming out, by a factor of 0.95. In other words, the electric power times 0.95 = 5222. By simple algebra, the power going in is 5194 / 0.95 = 5497 watts. The motor would be using 5497 watts if the power factor were 1. But the actual power factor is 0.8, so we need more apparent power going in order to get 5497 watts. That amount of power times 0.8 = 5497. Again, from simple algebra we can calculate that the apparent power is 5497 / 0.8 = 6874 watts. Finally, we can calculate the current:

$$I = P / E = 6874 / 120 = 57.283 \text{ amps.}$$

The closest answer on the test is 57.2, which is the correct answer.

The other problem is simpler:

A 3 horsepower, 100 V DC motor is 85% efficient when developing its rated output. What is the current?

This 3 horsepower motor is putting out 746 x 3 = 2238 watts of mechanical power. It's 85% efficient, which means that 2238 is 85% of the input power. Therefore, the input power is 2238 / .85 = 2633 watts. This is a DC motor, so we can immediately calculate the current:

$$I = P / E = 2633 / 100 = 26.33 \text{ amps.}$$

The closest answer on the test is 26.300, which is the correct answer.

Finally, you need to know that he output of a separately-excited AC generator running at a constant speed can be controlled by the amount of field current.

3-62I1 What occurs if the load is removed from an operating series DC motor?

A.   It will stop running.

B.   Speed will increase slightly.

C.   No change occurs.

D.   It will accelerate until it falls apart.

3-62I2 If a shunt motor running with a load has its shunt field opened, how would this affect the speed of the motor?

A.   It will slow down.

B.   It will stop suddenly.

C.   It will speed up.

D.   It will be unaffected.

3-62I3   The expression "voltage regulation" as it applies to a shunt-wound DC generator operating at a constant frequency refers to:

A.   Voltage fluctuations from load to no-load.

B.   Voltage output efficiency.

C.   Voltage in the secondary compared to the primary.

D.   Rotor winding voltage ratio

3-62I4   What is the line current of a 7 horsepower motor operating on 120 volts at full load, a power factor of 0.8, and 95% efficient?

A.   4.72 amps.

B.   13.03 amps.

C.   56 amps.

D. 57.2 amps.

3-6I5 A 3 horsepower, 100 V DC motor is 85% efficient when developing its rated output. What is the current?

A. 8.545 amps.

B. 20.345 amps.

C. 26.300 amps.

D. 25.000 amps.

3-6I6 The output of a separately-excited AC generator running at a constant speed can be controlled by:

A. The armature.

B. The amount of field current.

C. The brushes.

D. The exciter.

29: RMS AND PEP

There are a number of questions on the test regarding the various ways to measure AC (or RF) voltage and power. These questions might seem confusing, but if you read the questions carefully and follow the steps explained here, you will be able to answer all of them. On the other hand, if you discover that the math for this part of the test is out of your league, there are only a few questions, and you can simply memorize them.

The **RMS** (root mean square) voltage of an AC (or RF) signal is the value that results in the same power dissipation as a DC voltage of the same value. In other words, if we're worried about power, then RMS is the equivalent DC voltage. It is sometimes called the "effective" voltage. If you have a question involving power and AC voltage, then you will need to work with the RMS voltage, which might be different from the voltage value that is given.

Generally, an AC meter (such as an AC ammeter) will indicate RMS values. As you know, typical household voltage is 117 volts AC. This is the RMS value of the voltage.

The RMS voltage is about 0.7 times the peak voltage. To put it another way, the peak voltage is about 1.4 times the RMS voltage. (Instead of 1.4, it's really the square root of 2, but 1.4 is close enough.)

With this fact in mind, we can answer one of the questions on the test very easily: What is the peak voltage of a normal household outlet? We know that the RMS value is 117 volts. To get the peak voltage, we simply multiply by 1.4, for an answer of 163.8. Because of rounding errors, this is not the same as the value given on the test, but the closest answer, 165.5 volts, is the correct answer.

Be careful of the terminology, because "peak" voltage is not the same as "peak to peak". The peak voltage is the maximum positive or negative value that the sine wave reaches. The peak to peak is the difference between these two numbers—the distance from the very top of the graph to the very bottom of the graph. Therefore, peak-to-peak voltage is twice the peak voltage. Peak-to-peak voltage is the easiest voltage to measure on an oscilloscope, because you merely measure the distance from the very top to the very bottom.

The term **average voltage** is also sometimes used. This is defined as being 0.9 times the RMS value.

The following question is very straightforward: What would be the voltage across a 50-ohm dummy load dissipating 1200 watts? We are given watts and ohms, and we need to find volts. And since we're dealing with power, we need to think in terms of RMS voltage, which is what the question is asking for.

From the first chapter, we remember that watts equals volts squared divided by ohms. So we know that 1200 equals volts squared divided by 50:

$1200 = V^2 / 50$

We now think back to our high school algebra. We must solve the equation for V:

$1200 = V^2 / 50$

$1200 \times 50 = V^2$

$60000 = V^2$

Then, we take the square root of both sides of the equation, and get:

244.9 = V

The closest answer is 245, which is the correct answer.

The next problem is a bit trickier: What is the output PEP from a transmitter if an oscilloscope measures 200 volts peak-to-peak across a 50-ohm resistor connected to the transmitter output? Here, we know volts and ohms, and we need to calculate watts. We know that watts equals volts squared divided by ohms.

But before we use that equation, we need to change the voltage to RMS. Remember: whenever we're working with volts and trying to figure out power, we always need to work with RMS volts.

You need to read the question carefully, because we are starting out with peak-to-peak volts. This is twice the peak voltage. Since the peak-to-peak voltage is 200, the peak voltage is 100.

We know that RMS equals peak times 0.7. So the RMS voltage is 100 x 0.7 = 70.

Now that we have the RMS voltage, we can finally calculate the power. Watts equals volts squared divided by ohms. So here, watts equals 70 squared divided by 50, which equals 98. The closest answer is 100 watts, which is the correct answer.

The next problem works the same way: What is the output PEP from a transmitter if an oscilloscope measures 200 volts peak-to-peak across a 50-ohm dummy load connected to the transmitter output? Again, we have volts and ohms, and need to calculate watts.

We first change the 200 volts peak-to-peak to 100 watts peak. Then, we multiply by 0.7 to get 70 volts RMS. Then, we get the power as being volts squared divided by ohms. When we punch that into our calculator, we get an answer of 98 watts. The closest answer is 100 watts, which is the correct answer.

3-6A1 An AC ammeter indicates:

    A. Effective (TRM) values of current.

B. Effective (RMS) values of current.

C. Peak values of current.

D. Average values of current.

3-6A2 By what factor must the voltage of an AC circuit, as indicated on the scale of an AC voltmeter, be multiplied to obtain the peak voltage value?

A. 0.707

B. 0.9

C. 1.414

D. 3.14

3-6A3 What is the RMS voltage at a common household electrical power outlet?

A. 331-V AC.

B. 82.7-V AC.

C. 165.5-V AC.

D. 117-V AC.

3-6A4 What is the easiest voltage amplitude to measure by viewing a pure sine wave signal on an oscilloscope?

A. Peak-to-peak.

B. RMS.

C. Average.

D. DC.

3-6A5 By what factor must the voltage measured in an AC circuit, as indicated on the scale of an AC voltmeter, be multiplied to obtain the average voltage value?

A. 0.707

B. 1.414

C. 0.9

D. 3.14

3-6A6 What is the peak voltage at a common household electrical outlet?

A. 234 volts.

B. 117 volts.

C. 331 volts.

D. 165.5 volts.

3-8A1 What is the term used to identify an AC voltage that would cause the same heating in a resistor as a corresponding value of DC voltage?

A. Cosine voltage.

B. Power factor.

C. Root mean square (RMS).

D. Average voltage.

3-10B1 What is the peak-to-peak RF voltage on the 50 ohm output of a 100 watt transmitter?

A. 70 volts.   C. 140 volts.

B. 100 volts.   **D. 200 volts.**

3-12B4 What is the equivalent to the root-mean-square value of an AC voltage?

A. AC voltage is the square root of the average AC value.

B. The DC voltage causing the same heating in a given resistor at the peak AC voltage.

C. The AC voltage found by taking the square of the average value of the peak AC voltage.

D. The DC voltage causing the same heating in a given resistor as the RMS AC voltage of the same value.

3-12B5 What is the RMS value of a 340-volt peak-to-peak pure sine wave?

A. 170 volts AC.   **C. 120 volts AC.**

B. 240 volts AC.   D. 350 volts AC.

3-55G5 What is the output peak envelope power from a transmitter as measured on an oscilloscope showing 200 volts peak-to-peak across a 50-ohm load resistor?

A. 1,000 watts.

B. 100 watts.

C. 200 watts.

D. 400 watts.

3-55G6 What would be the voltage across a 50-ohm dummy load dissipating 1,200 watts?

A. 245 volts.

B. 692 volts.

C.   346 volts.

D.   173 volts.

30: SAFETY

The test covers a number of areas about safety. These questions are summed up by the following:

A lead-acid battery is potentially dangerous while being charged because of the emission of hydrogen gas. The hydrogen gas is dangerous because it is highly explosive.

There are two questions about eye protection. When repairing circuit board assemblies, it is most important to wear safety glasses. Don't be confused by this question—it doesn't look like a "safety" question, and the wrong answers all seem like reasonable things to do. But according to the FCC, wearing safety glasses is **the most important**. One question asks whether you should use a power drill without eye protection. The correct answer is "no". There are some tempting exceptions in the wrong answers, but according to the FCC, you should **never** use a drill without eye protection, not even in an extreme emergency.

When installing cables in a mobile unit, the most important factor is running the cables to that they don't interfere with the normal operation of the vehicle.

Prior to testing any RADAR system, you must assure that no personnel are in front of the antenna. This is because RF exposure can cause damage from RF heating of tissue, potentially causing cataracts.

You need to be aware that the U.S. Health Department limits exposure to microwave energy to 5.0 mW/centimeter.

The FCC also has Maximum Permissible Exposure (MPE) requirements. These limits take into account the average dose, so they are averaged over time. In "controlled" environments, they are averaged over 6 minutes. In "uncontrolled" environments, they are averaged over 30 minutes. Under FCC rules, an MPE study is required if the aggregate power level is 1000 Watts ERP. If the MPE power is present, then people working in the affected area need to be trained and certified yearly.

If you are working at a site with multiple transmitting antennas, then you must take into account all antennas with more than 5% of the maximum power density exposure limit.

You must never look into a fiber optic cable, because a signal could burn your retina, and if it is an infra-red signal, you would be unable to see it.

There are a few questions about lighting protection. First of all, it is impossible to protect against a direct lightning hit. But in the case of indirect hits, good grounding practices are usually critical. The main rule about grounding is that ground leads should run as straight as possible, without sharp corners. This is because lightning can jump off the ground lead when it reaches a sharp bend. The only method of lightning protection addressed on the test is the shorted-stub lightning protector. Since this is a stub tuned for a particular radio frequency, it works only at that tuned frequency.

There is one question about fire extinguishers. You need to know that FE30 is the type of fire extinguisher for electrical fires. You need to remember the number 30. Perhaps you can remember that EO stands for "electrical outlet", and that the 3 is an E backwards.

Finally, you need to know what a GFI electrical socket is. Electricians these days seem to call them GFCI's, but the FCC calls them GFI's. They prevent electric shock by sensing a ground path current and shutting down the circuit.

3-60I1 When a lead-acid storage battery is being charged, a harmful effect to humans is:

A.   Internal plate sulfation may occur under constant charging.

B.   Emission of oxygen.

C.   Emission of chlorine gas.

D.   Emission of hydrogen gas.

3-73K4 Before ground testing an aircraft RADAR, the operator should:

A.   Ensure that the area in front of the antenna is clear of other maintenance personnel to avoid radiation hazards.

B.   Be sure the receiver has been properly shielded and grounded.

C.   First test the transmitter connected to a matched load.

D.   Measure power supply voltages to prevent circuit damage.

3-79L6 When repairing circuit board assemblies it is most important to:

A.   Use a dental pick to clear plated-through holes.

B.   Bridge broken copper traces with solder.

C.   Wear safety glasses.

D.   Use a holding fixture.

3-80L2 What is most important when routing cables in a mobile unit?

A.   That cables be cut to the exact length.

B.   Assuring accessibility of the radio for servicing from outside the vehicle.

C.   Assuring radio or electronics cables do not interfere with the normal operation of the vehicle.

D.   Assuring cables are concealed under floor mats or carpeting.

3-94O2 Prior to testing any RADAR system, the operator should first:

A. Check the system grounds.

B. Assure the display unit is operating normally.

C. Inform the airport control tower or ship's master.

D. Assure no personnel are in front of the antenna.

3-94O5 Exposure to microwave energy from RADAR or other electronics devices is limited by U.S. Health Department regulations to _____ mW/centimeter.

A. 0.005

B. 5.0

C. 0.05

D. 0.5

3-99Q1 Compliance with MPE, or Maximum Permissible Exposure to RF levels (as defined in FCC Part 1, OET Bulletin 65) for "controlled" environments, are averaged over _____ minutes, while "uncontrolled" RF environments are averaged over _____ minutes.

A. 6, 30.

B. 30, 6.

C. 1, 15.

D. 15, 1.

3-99Q2 Sites having multiple transmitting antennas must include antennas with more than _____% of the maximum permissible power density exposure limit when evaluating RF site exposure.

A. Any

B. 5

C. 1

D. 12.5

3-99Q3 RF exposure from portable radio transceivers may be harmful to the eyes because:

A. Magnetic fields blur vision.

B. RF heating polarizes the eye lens.

C. The magnetic field may attract metal particles to the eye.

D. RF heating may cause cataracts.

3-99Q4 At what aggregate power level is an MPE (Maximum Permissible Exposure) study required?

A. 1000 Watts ERP.

B. 500 Watts ERP.

C. 100 Watts ERP.

D. Not required.

3-99Q5 Why must you never look directly into a fiber optic cable?

A. High power light waves can burn the skin surrounding the eye.

B. An active fiber signal may burn the retina and infra-red light cannot be seen.

C. The end is easy to break.

D. The signal is red and you can see it.

3-99Q6 If the MPE (Maximum Permissible Exposure) power is present, how often must the personnel accessing the affected area be trained and certified?

   A. Weekly.

   B. Monthly.

   C. Yearly.

   D. Not at all.

3-100Q1 What device can protect a transmitting station from a direct lightning hit?

   A. Lightning protector.

   B. Grounded cabinet.

   C. Short lead in.

   D. There is no device to protect a station from a direct hit from lightning.

3-100Q2 What is the purpose of not putting sharp corners on the ground leads within a building?

   A. No reason.

   B. It is easier to install.

   C. Lightning will jump off of the ground lead because it is not able to make sharp bends.

   D. Ground leads should always be made to look good in an installation, including the use of sharp bends in the corners.

3-100Q3 Should you use a power drill without eye protection?

   A. Yes.

B. No.

C. It's okay as long as you keep your face away from the drill bit.

D. Only in an extreme emergency.

3-100Q4 What class of fire is one that is caused by an electrical short circuit and what is the preferred substance used to extinguish that type of fire?

A. FE28.

B. FE29.

C. FE30.

D. FE31.

3-100Q5 Do shorted-stub lightning protectors work at all frequencies?

A. Yes.

B. No, the short also kills the radio signals.

C. No, the short enhances the radio signal at the tuned band.

D. No, only at the tuned frequency band.

3-100Q6 What is a GFI electrical socket used for?

A. To prevent electrical shock by sensing ground path current and shutting the circuit down.

B. As a gold plated socket.

C. To prevent children from sticking objects in the socket.

D. To increase the current capacity of the socket.

31: TTL

The test has a number of questions regarding TTL (transistor-transistor logic) digital logic gates.

A logic gate is a simple circuit that looks at two input voltages and provides an output voltage based upon a logical conclusion. You don't need to know the term for the test, but the relationships are known as Boolean logic.

There are three basic logic gates: AND, OR, and NOT. In an AND gate, if both inputs are "true", then the output is also "true". In other words, this type of gate answers the question, "are both of the inputs true"? If that's the case, then the answer to the question is also "true".

In an OR gate, if either input is true, then the output is also true. In other words, this type of gate answers the question, "is at least one of the inputs true?" If so, then the answer to that question is also "true".

A NOT gate simply switches the answer around. If the input is "true", then the output is "false". If the input is "false", then the output is "true".

You might ask, "is what true?" In the case of TTL, we are really asking, "is there five volts at that point?" Five volts is defined as "true", and zero volts is defined as "false". That one exception is a question that asks about a "negative-logic circuit". In that case, as the name implies, the high level (5 volts) represents "false" or a binary zero. But on all of the other questions, you should assume that positive logic is being used, and that 5 volts is the same as "true".

For the exam, you'll need to know the diagrams used for these gates:

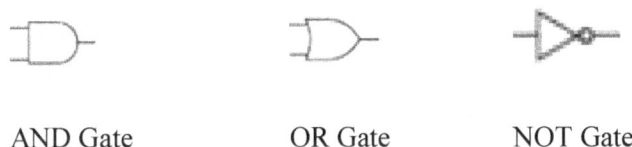

AND Gate   OR Gate   NOT Gate

One question also includes an XOR gate. An XOR gate answers the question, "is **exactly one** of the inputs true?" If one input is true and the other one is false, then the output is true. If both inputs are false, then the answer is false. The difference from an OR gate is that in an XOR gate, if both inputs are true, then the output is false. You don't need to know the name XOR for the test, but you do need to recognize the symbol:

XOR Gate

You also need to know the name of a NOR gate, which has a low output if any or all inputs are high. It is the same as an OR gate with a NOT gate connected to the output. A NAND gate is an AND gate with a NOT gate connected to the output. A chart showing input combinations and their corresponding outputs is called a truth table.

The exam will use the words "high" and "low" instead of "true" and "false". "High", meaning 5 volts, is the same as "true". In some questions, this is also referred to as a binary "1". "Low", meaning zero volts, is the same as "false". Some questions also refer to this is as a binary zero. With one exception, you can remember: High = True = 1 = 5 volts. Low = False = 0 = 0 volts. The only exception is one question that asks you about a "negative-logic circuit". As the name implies, in a negative-logic circuit, the voltages are reversed. The high voltage (5 volts) represents "false" or zero. The low voltage (zero volts) represents "true" or 1. In all of the other questions, you should assume that it is talking about normal positive logic.

One fact you do need to know is that if an input is not hooked up to anything (if it is left open), then it will have a tendency to get five volts at that input by itself. So if you see a connection on the test that is marked "open", you can assume that it is high.

Once you understand the definitions and symbols, the questions regarding logic gates are very simple logic problems. Let's look at one of the questions:

For the logic input levels shown in Figure 3E12, what are the logic levels of test points A, B and C in this circuit? (Assume positive logic.)

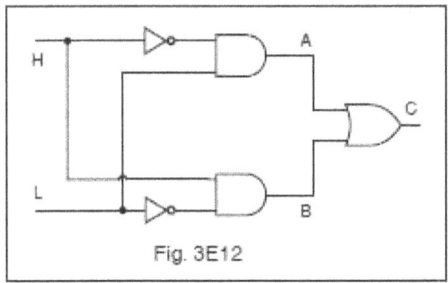

Fig. 3E12

Generally, you start at the left and work your way to the right. We'll start at the top input, which is marked H, meaning high. First, we can follow the top wire, which goes to a NOT gate. The NOT gate reverses the input, so the output is low. This goes into the AND gate at the top.

Now, let's start at the other input, which is marked L. Let's follow this wire, which splits and goes up to the other input of the top AND gate. This is connected directly to the input, and it is Low.

We now know the two inputs of the top AND gate. Both of them are low, or false. This AND gate is answering the question, "are BOTH of the inputs true?" The answer to that question is no, they are not both true. Therefore, the output of this gate is "false", or "low". We now know part of the answer, that A is low.

Let's start again with the top input of the circuit. This one is hooked directly to the bottom AND gate. Therefore, that input is high.

Now, let's trace the bottom input of the circuit as it goes to the right. It gets to a NOT gate. Since the input of the NOT gate is low, this gets reversed, and the output is high. This goes into the other input of the bottom AND gate.

We now know both inputs to the bottom AND gate. They are both high (true). And the AND gate answers the question, "are BOTH of the inputs true?" The answer to that question is yes, so the output of the gate is "true", or "high" We now know the next part of the answer, that B is high.

Finally, we have to figure out what the OR gate at the right does. Point A is low and point B is low. The OR gate answers the question, "is at least one of the inputs true?" The answer to this question is yes, so the output of the gate is "true" or "high". We now know the last part of the answer, that C is high.

As long as you follow this methodical approach, you will easily get all of these questions right.

Finally, you need to know what a logic probe is. A logic probe is simply a test instrument that lets you know if 5 volts is present. It's the best instrument for checking TTL circuits.

Speaking of binary, one question on the exam asks you how to write the number 2 using binary. As you probably know, the correct answer is 0010.

There's one final digital integrated circuit with which you need to be familiar for the exam. A decade counter is an IC that produces one output pulse for every ten input pulses.

3-33E4 TTL inputs left open develop what logic state?

A. A high-logic state.

B. A low-logic state.

C. Open inputs on a TTL device are ignored.

D. Random high- and low-logic states.

3-33E5 Which of the following instruments would be best for checking a TTL logic circuit?

    A. VOM.

    B. DMM.

    C. Continuity tester.

    D. Logic probe.

3-33E6 What do the initials TTL stand for?

    A. Resistor-transistor logic.

    B. Transistor-transistor logic.

    C. Diode-transistor logic.

    D. Emitter-coupled logic.

3-34E1 What is a characteristic of an AND gate?

    A. Produces a logic "0" at its output only if all inputs are logic "1".

    B. Produces a logic "1" at its output only if all inputs are logic "1".

    C. Produces a logic "1" at its output if only one input is a logic "1".

    D. Produces a logic "1" at its output if all inputs are logic "0".

3-34E2 What is a characteristic of a NAND gate?

    A. Produces a logic "0" at its output only when all inputs are logic "0".

    B. Produces a logic "1" at its output only when all inputs are logic "1".

    C. Produces a logic "0" at its output if some but not all of its inputs are logic "1".

D. Produces a logic "0" at its output only when all inputs are logic "1".

3-34E3 What is a characteristic of an OR gate?

A. Produces a logic "1" at its output if any input is logic "1".

B. Produces a logic "0" at its output if any input is logic "1".

C. Produces a logic "0" at its output if all inputs are logic "1".

D. Produces a logic "1" at its output if all inputs are logic "0".

3-34E4 What is a characteristic of a NOR gate?

A. Produces a logic "0" at its output only if all inputs are logic "0".

B. Produces a logic "1" at its output only if all inputs are logic "1".

C. Produces a logic "0" at its output if any or all inputs are logic "1".

D. Produces a logic "1" at its output if some but not all of its inputs are logic "1".

3-34E5 What is a characteristic of a NOT gate?

A. Does not allow data transmission when its input is high.

B. Produces a logic "0" at its output when the input is logic "1" and vice versa.

C. Allows data transmission only when its input is high.

D. Produces a logic "1" at its output when the input is logic "1" and vice versa.

3-34E6 Which of the following logic gates will provide an active high out when both inputs are active high?

A. NAND.

B. NOR.

C. AND.

D. XOR.

3-35E1 In a negative-logic circuit, what level is used to represent a logic 0?

A. Low level.

B. Positive-transition level.

C. Negative-transition level.

D. High level.

3-35E2 For the logic input levels shown in Figure 3E12, what are the logic levels of test points A, B and C in this circuit? (Assume positive logic.)

A. A is high, B is low and C is low.

B. A is low, B is high and C is high.

C. A is high, B is high and C is low.

D. A is low, B is high and C is low.

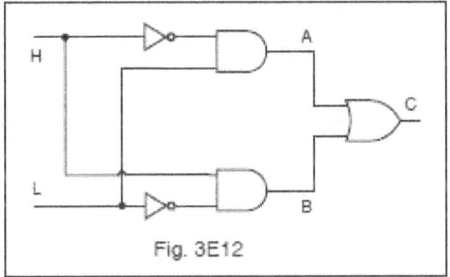

Fig. 3E12

3-35E3 For the logic input levels given in Figure 3E13, what are the logic levels of test points A, B and C in this circuit? (Assume positive logic.)

A. A is low, B is low and C is high.

B. A is low, B is high and C is low.

C. A is high, B is high and C is high.

D. A is high, B is low and C is low.

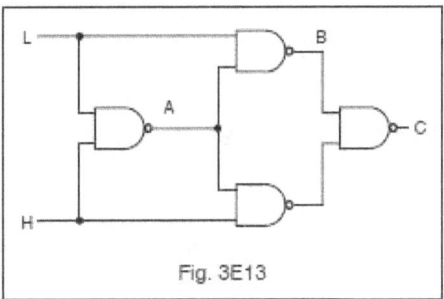

Fig. 3E13

3-35E4 In a positive-logic circuit, what level is used to represent a logic 1?

A. High level

B. Low level

C. Positive-transition level

D. Negative-transition level

3-35E5 Given the input levels shown in Figure 3E14 and assuming positive logic devices, what would the output be?

A. A is low, B is high and C is high.

B. A is high, B is high and C is low.

C. A is low, B is low and C is high.

D. None of the above are correct.

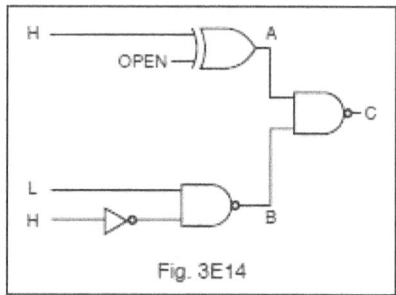

Fig. 3E14

3-35E6 What is a truth table?

A. A list of input combinations and their corresponding outputs that characterizes a digital device's function.

B. A table of logic symbols that indicate the high logic states of an op-amp.

C. A diagram showing logic states when the digital device's output is true.

D. A table of logic symbols that indicates the low logic states of an op-amp.

3-40E3 What is the function of a decade counter digital IC?

A. Decode a decimal number for display on a seven-segment LED display.

B. **Produce one output pulse for every ten input pulses.**

C.   Produce ten output pulses for every input pulse.

D.   Add two decimal numbers.

3-40E6   In binary numbers, how would you note the quantity TWO?

A. **0010**

B.   0002

C.   2000

D.   0020

3-74L5   What instrument is used to indicate high and low digital voltage states?

A.   Ohmmeter.

B. **Logic probe.**

C.   Megger.

D.   Signal strength meter.

32: MULTIVIBRATORS AND FLIP-FLOPS

A multivibrator is the general term for any circuit with two states. They can be monostable, bistable, unstable, or astable. For the exam, you need to know only the following facts:

An astable multivibrator alternates between the two unstable states. The output of an astable multivibrator is a square wave.

A monostable multivibrator can be switch momentarily to the opposite binary state and then returns after a set time.

A bistable multivibrator is also known as a flip-flop. It can be used to store one bit of data. Since each flip-flop stores one bit of data, one easy question asks you how many are needed to store 8 bits. The answer, of course, is 8.

You don't need to know this for the exam, but it consists of two NOR gates. It is a type of multivibrator circuit, which is the general term for any circuit with two states. A flip-flop, therefore, has two stable states (on and off).

A flip-flop (referred to in one question by its other name, bistable multivibrator, can be used as a frequency divider. For the exam, you need to know that to divide a signal frequency by 4, you would need 2 flip-flops.

Finally, you need to know that a flip-flop can not be used to operate in toggle mode with R-S inputs held constant and CLK initiated. Just remember, for the test, that "toggle mode" is the one listed answer that cannot be done.

3-36E1 A flip-flop circuit is a binary logic element with how many stable states?

    A. 1

    B. 2

    C. 4

    D. 8

3-36E2 What is a flip-flop circuit? A binary sequential logic element with ___ stable states.

    A. 1

    B. 4

    C. 2

    D. 8

3-36E3 How many flip-flops are required to divide a signal frequency by 4?

    A. 1

    B. 4

    C. 8

    D. 2

3-36E4 How many bits of information can be stored in a single flip-flop circuit?

A. 1

B. 2

C. 3

D. 4

3-36E5 How many R-S flip-flops would be required to construct an 8 bit storage register?

A. 2

B. 4

C. 8

D. 16

3-36E6 An R-S flip-flop is capable of doing all of the following except:

A. Accept data input into R-S inputs with CLK initiated.

B. Accept data input into PRE and CLR inputs without CLK being initiated.

C. Refuse to accept synchronous data if asynchronous data is being input at same time.

D. Operate in toggle mode with R-S inputs held constant and CLK initiated.

3-37E1 The frequency of an AC signal can be divided electronically by what type of digital circuit?

A. Free-running multivibrator.

B. Bistable multivibrator.

C. OR gate.

D.   Astable multivibrator.

3-37E2   What is an astable multivibrator?

A.   A circuit that alternates between two stable states.

B.   A circuit that alternates between a stable state and an unstable state.

C.   A circuit set to block either a 0 pulse or a 1 pulse and pass the other.

D.   A circuit that alternates between two unstable states.

3-37E3   What is a monostable multivibrator?

A.   A circuit that can be switched momentarily to the opposite binary state and then returns after a set time to its original state.

B.   A "clock" circuit that produces a continuous square wave oscillating between 1 and 0.

C.   A circuit designed to store one bit of data in either the 0 or the 1 configuration.

D.   A circuit that maintains a constant output voltage, regardless of variations in the input voltage.

3-37E4   What is a bistable multivibrator circuit commonly named?

A.   AND gate.

B.   OR gate.

C.   Clock.

D.   Flip-flop.

3-37E5   What is a bistable multivibrator circuit?

A. Flip-flop.

B. AND gate.

C. OR gate.

D. Clock.

3-37E6 What wave form would appear on the voltage outputs at the collectors of an astable, multivibrator, common-emitter stage?

A. Sine wave.

B. Sawtooth wave.

C. Square wave.

D. Half-wave pulses.

33: MICROPROCESSORS AND MEMORY

The exam contains a number of quite simple questions regarding memory and microprocessors. You probably are already familiar with most of these concepts, and the few you don't know already can be memorized quite easily. Most merely ask you to know the names of various components. One question, for example, asks you to name the type of memory that can be accessed randomly. As you probably guessed, the answer is "random access memory" (RAM).

DRAM (dynamic random-access memory) is defined as the random-accessed memory that must be refreshed periodically.

ROM (read-only memory) is defined as a fixed pattern of digital data stored in a semiconductor's memory matrix.

"IO" stands for input-output. A program is defined as a microprocessor's sequence of commands and instructions.

"BCD" stands for "binary coded decimal".

"ALU" stands for "arithmetical logic unit."

A "watchdog timer" in a computer-controlled radio verifies that the microprocessor is executing the program.

The "clock" in a microprocessor generates pulses.

The data bus line connects the microprocessor with the memory and inputs and outputs.

You also need to know the following two facts: If a memory IC has 4 data bus input/output pins and 4 address pins, then it contains 64 individual memory cells. The MCU processor does NOT contain a voltage regulator. (As you probably could have guessed, it contains only digital components.)

3-38E1 What is the name of the semiconductor memory IC whose digital data can be written or read, and whose memory word address can be accessed randomly?

   A. ROM – Read-Only Memory.

   B. PROM – Programmable Read-Only Memory.

   C. RAM – Random-Access Memory.

   D. EPROM – Electrically Programmable Read-Only Memory.

3-38E2 What is the name of the semiconductor IC that has a fixed pattern of digital data stored in its memory matrix?

   A. RAM – Random-Access Memory.

   B. ROM – Read-Only Memory.

   C. Register.

   D. Latch.

3-38E3 What does the term "IO" mean within a microprocessor system?

   A. Integrated oscillator.

   B. Integer operation.

   C. Input-output.

D   Internal operation.

3-38E4 What is the name for a microprocessor's sequence of commands and instructions?

A. Program.

B. Sequence.

C. Data string.

D. Data execution.

3-38E5 How many individual memory cells would be contained in a memory IC that has 4 data bus input/output pins and 4 address pins for connection to the address bus?

A. 8

B. 16

C. 32

D. 64

3-38E6 What is the name of the random-accessed semiconductor memory IC that must be refreshed periodically to maintain reliable data storage in its memory matrix?

A. ROM – Read-Only Memory.

B. DRAM – Dynamic Random-Access Memory.

C. PROM – Programmable Read-Only Memory.

D. PRAM – Programmable Random-Access Memory.

3-39E1 In a microprocessor-controlled two-way radio, a "watchdog" timer:

A. **Verifies that the microprocessor is executing the program.**

B. Assures that the transmission is exactly on frequency.

C. Prevents the transmitter from exceeding allowed power out.

D. Connects to the system RADAR presentation.

3-39E3 Which of the following is not part of a MCU processor?

A. RAM

B. ROM

C. I/O

D. **Voltage Regulator**

3-39E4 What portion of a microprocessor circuit is the pulse generator?

A. **Clock**

B. RAM

C. ROM

D. PLL

3-39E5 In a microprocessor, what is the meaning of the term "ALU"?

A. Automatic lock/unlock.

B. **Arithmetical logic unit.**

C. Auto latch undo.

D. Answer local unit.

3-39E6 What circuit interconnects the microprocessor with the memory and input/output system?

A. Control logic bus.

B. PLL line.

C. Data bus line.

D. Directional coupler.

3-40E2 What does the term "BCD" mean?

A. Binaural coded digit.

B. Bit count decimal.

C. Binary coded decimal.

D. Broad course digit.

34: AIR NAVIGATION AND COMMUNICATION

The exam includes a number of questions regarding aircraft navigational equipment.

VOR (VHF omnidirectional range) is a short-range navigation system for aircraft. VOR operates on 108-117.95 MHz. The ground station sends out two signals. One of these signals is transmitted in all directions. The other signal is transmitted through a directional antenna rotating 30 times per second (or the electronic equivalent). The aircraft receiver measures the phase angle between the two signals, and can determine the "radial" on which it lies--it other words, the direction the rotating antenna is pointing at the time the strongest signal is received. The two signals are in phase with each other when the antenna is pointing magnetic north (360 degrees). (One of the questions on the test uses the word "azimuth", which is really just another word for direction.) While VOR might be phased out in the future, it is still the standard air navigation system worldwide. As with airport runways, all directions associated with VOR are related to magnetic north rather than true north. The aircraft VOR antenna is a horizontally polarized omnidirectional antenna.

DME (distance measuring equipment) is often mentioned in reference to VOR, because DME stations are often (but not always) at the same location as VOR stations. (On one question, the combined ground station is referred to as the VORTAC station.) As the name implies, DME is used to measure the distance from the aircraft to the station. The aircraft sends a signal which interrogates the ground station, and the ground station replies with an identical pulse sequence. As with RADAR, the aircraft DME measures the time and calculates the distance to the station. The actual measurement is called the slant range, because it is the direct distance between the aircraft and the receiver, which is greater than the distance along the ground. The DME station has a delay before responding to the interrogation (typically, 50 microseconds). The purpose of this delay is to allow operation at close range. (If it weren't for this delay, the aircraft DME wouldn't have time to switch from transmit to receive if the distance was too close.)

ILS (instrument landing system) is a system used for landing an aircraft. The transmitter (108.1-111.95 MHz) transmitter transmits two signals from directional antennas located at the end of the runway. There are two antennas, one for the localizer (LOC), which tells the aircraft whether it is in the correct horizontal position, or if it is too far to the right or left of the runway. The other, glide slope, tells the aircraft whether it is too high or too low. The aircraft merely follows the center line to a safe landing.

The aircraft determines this from the fact that the modulation differs from right to left and from high to low. The modulation is at 90 Hz and 150 Hz. 90 Hz indicates that the aircraft is too far to the left, or too high. 150 Hz indicates that the aircraft is too far to the right, or too low. If the modulation is equal at both frequencies, this means that the aircraft is on the desired path.

For the exam, you need to know that most DME systems automatically tune their transmitter and receiver to the paired VOR/LOC channel.

An aircraft uses two antennas for the ILS. The glide scope antenna is typically a folded dipole, and the marker beacon antenna is a balanced loop.

ADF is an older system of navigation which is still widely used by both aircraft and boats. It uses LF or MF ground stations that are transmitting a continuous beacon. These stations can be non directional beacon (NDB) stations which are intended for navigation and shown on charts. They can also be standard AM broadcast stations, as long as the pilot is aware of their location. (In fact, the transmitter locations of many AM stations are shown on charts for this reason.) Because the ground stations transmit on LF and MF frequencies, the signals can follow the curvature of the earth and can be used for navigation hundreds of miles away. ADF is less reliable at night, due to the "night effect": At night, beacon signals can bounce off the ionosphere, and can be received at almost any direction.

There are a few questions on the test regarding aircraft RADAR transponder. Most aircraft contain a responder. When the aircraft is picked up by ground RADAR, the transponder sends information about the aircraft, which is displayed on the RADAR console. As mentioned previously, these systems use the frequencies 1030 and 1090 MHz. The ground station transmits on 1030 MHz, and the aircraft transmits on 1090 MHz. The ground station is transmitting at 1030 MHz in a directional pattern, in conjunction with the directional RADAR antenna. The aircraft is transmitting pulsed data on 1090 MHz. The aircraft antenna for the transponder is an omnidirectional L-band monopole. A mode S transponder is an enhanced version. The mode S transponder also has mid-air collision avoidance capabilities.

There are only a few questions on the test that relate specifically to aircraft communications. You need to know:

1. A typical aircraft HF transmitter is 100 watts, and a typical VHF transmitter is 20 watts.
2. Aviation communications generally uses dynamic microphones.
3. SELCAL (selective calling), as the name implies, is a system where a ground-based transmitter can call a selected aircraft without the crew continuously monitoring that frequency.

3-68K2 The Distance Measuring Equipment (DME) measures the distance from the aircraft to the DME ground station. This is referred to as:

A. DME bearing.

B. The slant range.

C. Glide Slope angle of approach.

D. Localizer course width.

3-68K3 The Distance Measuring Equipment (DME) ground station has a built-in delay between reception of an interrogation and transmission of the reply to allow:

A. Someone to answer the call.

B. The VOR to make a mechanical hook-up.

C. Operation at close range.

D. Clear other traffic for a reply.

3-68K4 What is the main underlying operating principle of an aircraft's Distance Measuring Equipment (DME)?

A. A measurable amount of time is required to send and receive a radio signal through the Earth's atmosphere.

B. The difference between the peak values of two DC voltages may be used to determine an aircraft's distance to another aircraft.

C. A measurable frequency compression of an AC signal may be used to determine an aircraft's altitude above the earth.

D. A phase inversion between two AC voltages may be used to determine an aircraft's distance to the exit ramp of an airport's runway.

3-68K5 What radio navigation aid determines the distance from an aircraft to a selected VORTAC station by measuring the length of time the radio signal takes to travel to and from the station?

A. RADAR.

B. Loran C.

C. Distance Marking (DM).

D. Distance Measuring Equipment (DME).

3-68K6 The majority of airborne Distance Measuring Equipment systems automatically tune their transmitter and receiver frequencies to the paired __ / __ channel.

    A. VOR/marker beacon.

    B. VOR/LOC.

    C. Marker beacon/glideslope.

    D. LOC/glideslope.

3-69K1 All directions associated with a VOR station are related to:

    A. Magnetic north.

    B. North pole.

    C. North star.

    D. None of these.

3-69K2 The rate that the transmitted VOR variable signal rotates is equivalent to how many revolutions per second?

    A. 60

    B. 30

    C. 2400

    D. 1800

3-69K4 Lines drawn from the VOR station in a particular magnetic direction are:

    A. Radials.

    B. Quadrants.

    C. Bearings.

D.   Headings.

3-69K5 The amplitude modulated variable phase signal and the
frequency modulated reference phase signal of a Very-high-frequency
Omnidirectional Range (VOR) station used for aircraft navigation are
synchronized so that both signals are in phase with each other at
_____ of the VOR station.

A.   180 degrees South, true bearing position.

B.   360 degrees North, magnetic bearing position.

C.   180 degrees South, magnetic bearing position.

D.   0 degrees North, true bearing position.

3-69K6 What is the main underlying operating principle of the Very-
high-frequency Omnidirectional Range (VOR) aircraft navigational
system?

A.   A definite amount of time is required to send and receive a radio
signal.

B.   The difference between the peak values of two DC voltages may
be used to determine an aircraft's altitude above a selected VOR station.

**C.   A phase difference between two AC voltages may be used to
determine an aircraft's azimuth position in relation to a selected
VOR station.**

D.   A phase difference between two AC voltages may be used to
determine an aircraft's distance from a selected VOR station.

3-70K3 Which of the following is a required component of an
Instrument Landing System (ILS)?

A.   Altimeter: shows aircraft height above sea-level.

**B.   Localizer: shows aircraft deviation horizontally from center
of runway.**

C.   VHF Communications: provide communications to aircraft.

D.   Distance Measuring Equipment: shows aircraft distance to VORTAC station.

3-70K4 What type of antenna is used in an aircraft's Instrument Landing System (ILS) glideslope installation?

A.   A vertically polarized antenna that radiates an omnidirectional antenna pattern.

B.   A balanced loop reception antenna.

C.   A folded dipole reception antenna.

D.   An electronically steerable phased-array antenna that radiates a directional antenna pattern.

3-70K5 Choose the only correct statement about the localizer beam system used by aircraft to find the centerline of a runway during an Instrument Landing System (ILS) approach to an airport. The localizer beam system:

A.   Operates within the assigned frequency range of 108.10 to 111.95 GHz.

B.   Produces two amplitude modulated antenna patterns; one pattern above and one pattern below the normal 2.5 degree approach glide path of the aircraft.

C.   Frequencies are automatically tuned-in when the proper glide slope frequency is selected on the aircraft's Navigation and Communication (NAV/COMM) transceiver.

D.   Produces two amplitude modulated antenna patterns; one pattern with an audio frequency of 90 Hz and one pattern with an audio frequency of 150 Hz, one left of the runway centerline and one right of the runway centerline.

3-70K6 On runway approach, an ILS Localizer shows:

A. Deviation left or right of runway center line.

B. Deviation up and down from ground speed.

C. Deviation percentage from authorized ground speed.

D. Wind speed along runway.

3-71K2 What is meant by the term "night effect" when using an aircraft's Automatic Direction Finding (ADF) equipment? Night effect refers to the fact that:

A. All Non Directional Beacon (NDB) transmitters are turned-off at dusk and turned-on at dawn.

B. Non Directional Beacon (NDB) transmissions can bounce-off the Earth's ionosphere at night and be received at almost any direction.

C. An aircraft's ADF transmissions will be slowed at night due to the increased density of the Earth's atmosphere after sunset.

D. An aircraft's ADF antennas usually collect dew moisture after sunset which decreases their effective reception distance from an NDB transmitter.

3-71K4 In addition to duplicating the functions of a mode C transponder, an aircraft's mode S transponder can also provide:

A. Primary RADAR surveillance capabilities.

B. Long range lightning detection.

C. Mid-Air collision avoidance capabilities.

D. Backup VHF voice communication abilities.

3-71K5 What type of encoding is used in an aircraft's mode C transponder transmission to a ground station of the Air Traffic Control RADAR Beacon System (ATCRBS)?

A. Differential phase shift keying.

B. Pulse position modulation.

C. Doppler effect compressional encryption.

D. Amplitude modulation at 95%.

3-71K6 Choose the only correct statement about an aircraft's Automatic Direction Finding (ADF) equipment.

A. An aircraft's ADF transmission exhibits primarily a line-of-sight range to the ground-based target station and will not follow the curvature of the Earth.

B. Only a single omnidirectional sense antenna is required to receive an NDB transmission and process the signal to calculate the aircraft's bearing to the selected ground station.

C. All frequencies in the ADF's operating range except the commercial standard broadcast stations (550 to 1660 kHz) can be utilized as a navigational Non Directional Beacon (NDB) signal.

D. An aircraft's ADF antennas can receive transmissions that are over the Earth's horizon (sometimes several hundred miles away) since these signals will follow the curvature of the Earth.

3-72K1 What type of antenna pattern is radiated from a ground station phased-array directional antenna when transmitting the PPM pulses in a Mode S interrogation signal of an aircraft's Traffic alert and Collision Avoidance System (TCAS) installation?

A. 1090 MHz directional pattern.

B. 1030 MHz omnidirectional pattern.

C. 1090 MHz omnidirectional pattern.

D. 1030 MHz directional pattern.

3-72K2 What type of antenna is used in an aircraft's Instrument Landing

System (ILS) marker beacon installation?

A. An electronically steerable phased-array antenna that radiates a directional antenna pattern.

B. A folded dipole reception antenna.

C. A balanced loop reception antenna.

D. A horizontally polarized antenna that radiates an omnidirectional antenna pattern.

3-72K6 What type of antenna is attached to an aircraft's Mode C transponder installation and used to receive 1030 MHz interrogation signals from the Air Traffic Control Radar Beacon System (ATCRBS)?

A. An electronically steerable phased-array directional antenna.

B. An L-band monopole blade-type omnidirectional antenna.

C. A folded dipole reception antenna.

D. An internally mounted, mechanically rotatable loop antenna.

3-73K2 Aviation services use predominantly _____ microphones.

A. Dynamic

B. Carbon

C. Condenser

D. Piezoelectric crystal

3-73K3 Typical airborne HF transmitters usually provide a nominal RF power output to the antenna of _____ watts, compared with _____ watts RF output from a typical VHF transmitter.

A. 10, 50

B. 50, 10

C. 20, 100

D. 100, 20

3-73K5 What type of antenna is used in an aircraft's Very High Frequency Omnidirectional Range (VOR) and Localizer (LOC) installations?

A. Vertically polarized antenna that radiates an omnidirectional antenna pattern.

B. Horizontally polarized omnidirection reception antenna.

C. Balanced loop transmission antenna.

D. Folded dipole reception antenna.

3-73K6 What is the function of a commercial aircraft's SELCAL installation? SELCAL is a type of aircraft communications _____.

A. Device that allows an aircraft's receiver to be continuously calibrated for signal selectivity.

B. System where a ground-based transmitter can call a selected aircraft or group of aircraft without the flight crew monitoring the ground-station frequency.

C. Transmission that uses sequential logic algorithm encryption to prevent public "eavesdropping" of crucial aircraft flight data.

D. System where an airborne transmitter can selectively calculate the line-of-sight distance to several ground-station receivers.

35: MARINE COMMUNICATIONS

The exam covers quite a few marine communications systems. Even if you are unfamiliar with marine electronics, most of these questions are quite simple once you understand a few basic definitions and concepts.

There are only a few facts you need to know about VHF marine radio. All of these were covered in previous chapters, but to summarize: 25 kHz channel spacing is used with 5 kHz deviation. Channel 16 is used for distress and calling, and channel 70 is used for digital selective calling. For the emergency signaling functions to operate, the 9-digit MMSI must be entered.

GMDSS stands for Global Maritime Distress Safety System, and it is a set of safety procedures, types of equipment, and communications protocols used for marine safety and rescue. The components of GMDSS are: EPIRB's (emergency position-indicating radio beacons), NAVTEX, INMARSAT, HF radio, search and rescue locating devices, and digital selective calling. These are all covered on the test, with varying detail. Some of these have been addressed in other chapters, and we'll discuss the other points here.

An **EPIRB** is a 406 MHz emergency transmitter monitored by the COSPAS-SARSAT satellite system. In an emergency, the mariner activates the transmitter to summon help. The transmitter sends a unique hexadecimal identification number, which should have been previously registered with the agency responsible for the rescue. The satellite is able to determine the location of the transmitter. Therefore, the EPIRB should be able to alert rescuers who is in distress, and where they are.

The next element of GDMSS is **NAVTEX**, which stands for navigation text. This is a system of text bulletins transmitted on MF (usually 518 kHz) containing MSI (marine safety information) such as weather forecasts and bulletins. The receiver can be programmed to print only messages of interest to the particular ship. Each message begins with a subject indicator, and the receiver can be programmed to reject messages that are not relevant, such as weather in other parts of the world. NAVTEX uses SITOR-B or FEC mode of transmission.

SITOR stands for simplex teletype over radio. SITOR can be used for one-way communications to either a single station, or to all stations. In the case of NAVTEX, of course, it is being used for one-way communication to all ships. Therefore, as one of the questions points out, it is not necessary to be in two-way communication with the coast station in order to receive NAVTEX.

SITOR can also be used for two-way communications. For two-way communications, it can accomodate ARQ (automatic repeat request) error correction, which is made in data groups consisting of three-character blocks. For one-way communications to all stations (such as NAVTEX), it uses FEC (forward error correction). For one-way communications to a single station, it uses SFEC (selective FEC).

You need to know only two items regarding the SITOR and FEC protocols: Each SITOR character consists of 7 bits. Four of these bits are always zeros, and the other 3 are always ones. For FEC, you need to know that code 0000111 is phasing signal 1. If you memorize the number 0000111, you'll get both of these questions correct. You'll be able to count the number of ones and zeroes, in case you get the SITOR question. If you get the FEC question, simply look for that number as being the right answer.

The next elements of GDMSS are the INMARSAT communications satellite system and HF radio. The questions regarding these subjects are covered in other chapters.

The next element of GDMSS is the search and rescue transmitter (**SART**). The SART is a device carried in a lifeboat. It monitors for RADAR signals from other ships, and upon detecting one, sends out a signal. This signal shows up on the RADAR screen as a series of 12 equally spaced dots.

The final element of GDMSS is digital selective calling (DSC) on the VHF radio. As noted above, if the radio is equipped with DSC, then the 9-digit MMSI should be entered in case this feature is needed in an emergency. The code used for DSC transmissions consists of a 10 bit error correcting code starting with bits of data followed by a 3 bit error correcting code.

Finally, there are two questions regarding facsimile. You should know that the standard for HF weather facsimile transmissions is 120 lines per minute. You probably already know the definition of facsimile: "the transmission of printed pictures for permanent display on paper." This question is one case where too much knowledge might cause you to select the wrong answer. One wrong answer defines facsimile as "transmission of still pictures by slow-scan television." You might be tempted to select this answer, since a radio signal carrying SSTV could be identical to one carrying FAX. But that is the wrong answer! Historically, the FCC has treated television and facsimile as being two completely separate things. And by definition, the thing that makes it facsimile, and not television, is the fact that it's intended to be printed on paper. Therefore, it is important to remember that facsimile always means that there will be a printed picture on paper.

3-84M3 Which statement best describes the code used for GMDSS-DSC transmissions?

A. A 10 bit error correcting code starting with bits of data followed by a 3 bit error correcting code.

B. A 10 bit error correcting code starting with a 3 bit error correcting code followed by 7 bits of data.

C. An 8 bit code with 7 bits of data followed by a single parity bit.

D. A 7 bit code that is transmitted twice for error correction.

3-84M4 Which is the code used for SITOR-A and -B transmissions?

A. The 5 bit baudot telex code.

B.   Each character consists of 7 bits with 3 "zeros" and 4 "ones".

C.   Each character consists of 7 bits with 4 "zeros" and 3 "ones".

D.   Each character has 7 bits of data and 3 bits for error correction.

3-84M5 Which of the following statements is true?

A.   The Signal Repetition character (1001100) is used as a control signal in SITOR-ARQ.

B.   The Idle Signal (a) (0000111) is used for FEC Phasing Signal 1.

C.   The Idle Signal (b) (0011001) is used for FEC Phasing Signal 2.

D.   The Control Signal 1 (0101100) is used to determine the time displacement in SITOR-B.

3-86N4 Which of the following statements concerning SITOR communications is true?

A. ARQ message transmissions are made in data groups consisting of three-character blocks.

B.   ARQ transmissions are acknowledged by the Information Receiving Station only at the end of the message.

C.   ARQ communications rely upon error correction by time diversity transmission and reception.

D.   Forward error correction is an interactive mode.

3-86N5 The sequence ARQ, FEC, SFEC best corresponds to which of the following sequences?

A.   One-way communications to a single station, one-way communications to all stations, two-way communications.

B.   One-way communications to all stations, two-way communications, one-way communications to a single station.

C.   Two way communications, one-way communications to all stations, one-way communications to a single station.

D.   Two way communications, one-way communications to a single station, one-way communications to all stations.

3-86N6 Which of the following statements concerning SITOR communications is true?

A.   Communication is established on the working channel and answerbacks are exchanged before FEC broadcasts can be received.

B.   In the ARQ mode each character is transmitted twice.

C.   Weather broadcasts cannot be made in FEC because sending each character twice would cause the broadcast to be prohibitively long.

D.   Two-way communication with the coast radio station using FEC is not necessary to be able to receive the broadcasts.

3-87N1 What causes the SART to begin a transmission?

A.   When activated manually, it begins radiating immediately.

B.   After being activated the SART responds to RADAR interrogation.

C.   It is either manually or water activated before radiating.

D.   It begins radiating only when keyed by the operator.

3-87N2 How should the signal from a Search And Rescue RADAR Transponder appear on a RADAR display?

A.   A series of dashes.

B.   A series of spirals all originating from the range and bearing of the SART.

    C. A series of twenty dashes.

    D. A series of 12 equally spaced dots.

3-87N4 Which piece of required GMDSS equipment is the primary source of transmitting locating signals?

    A. Radio Direction Finder (RDF).

    B. A SART transmitting on 406 MHz.

    C. Survival Craft Transceiver.

    D. An EPIRB transmitting on 406 MHz.

3-87N5 Which of the following statements concerning satellite EPIRBs is true?

    A. Once activated, these EPIRBs transmit a signal for use in identifying the vessel and for determining the position of the beacon.

    B. The coded signal identifies the nature of the distress situation.

    C. The coded signal only identifies the vessel's name and port of registry.

    D. If the GMDSS Radio Operator does not program the EPIRB, it will transmit default information such as the follow-on communications frequency and mode.

3-87N6 What statement is true regarding 406 MHz EPIRB transmissions?

    A. Allows immediate voice communications with the RCC.

    B. Coding permits the SAR authorities to know if manually or automatically activated.

C. Transmits a unique hexadecimal identification number.

D. Radio Operator programs an I.D. into the SART immediately prior to activation.

3-88N1 What is facsimile?

A. The transmission of still pictures by slow-scan television.

B. The transmission of characters by radioteletype that form a picture when printed.

C. The transmission of printed pictures for permanent display on paper.

D. The transmission of video by television.

3-88N2 What is the standard scan rate for high-frequency 3 MHz - 23 MHz weather facsimile reception from shore stations?

A. 240 lines per minute.

B. 120 lines per minute.

C. 150 lines per second.

D. 60 lines per second.

3-88N4 Which of the following statements about NAVTEX is true?

A. Receives MSI broadcasts using SITOR-B or FEC mode.

B. The ship station transmits on 518 kHz.

C. The ship receives MSI broadcasts using SITOR-A or ARQ mode.

D. NAVTEX is received on 2182 kHz using SSB.

3-88N6 What determines whether a NAVTEX receiver does not print a particular type of message content?

A. The message does not concern your vessel.

B. The subject indicator matches that programmed for rejection by the operator.

C. The transmitting station ID covering your area has not been programmed for rejection by the operator.

D. All messages sent during each broadcast are printed.

36: RADAR

The GROL exam contains a number of questions about RADAR. Even if you have no experience with RADAR, you can easily answer most of these questions if you have a bit of basic knowledge and apply some common sense.

As you probably know, RADAR works by sending out a radio signal, having it bounce off distant objects, and then observing the reflected signal. To determine the direction, you simply have to know what direction the antenna was pointing. The **synchro transmitter and receiver** are used to transmit the antenna's angular position to the indicator. To determine the distance, you measure the time between sending out the signal and receiving the echo. Since the speed of light is known, you can calculate the distance.

For many of the problems, you need to know that the speed of light is 6.2 microseconds per nautical mile. You should memorize that number, since it will make most of the problems trivially easy. But should you forget, you can still answer most of the questions by getting out your calculator and scratch paper, as long as you know the speed of light (186,000 miles per second, or 300 million meters per second). Even if you don't know the exact length of a nautical mile (1.15 statute miles), you'll get most of the problems right if you remember only that the nautical mile is slightly longer than a mile. But the exam will be easier if you remember that a radar signal takes 6.2 microseconds to travel one nautical mile.

In other words, if the radar sends out a signal one nautical mile away, it will take 6.2 microseconds for the signal to reach the object. After it bounces off the object, it will take another 6.2 microseconds to return.

The total time elapsed will be 12.4 microseconds. (The exact value will be 12.346 µs.)

Most shipboard radar uses frequencies in the SHF portion of the spectrum. A radar sends out pulses. The length of the individual pulses will vary, depending upon the radar, and depending upon the settings of the radar. The number of pulses per second will also vary. The power output of a RADAR is provided by the **magnetron** tube, and the pulse rate of the radar is set by the pulse rate of the magnetron. You should memorize the following facts:

1. The normal range of pulse repetition rates is 500 to 2000 pulses per second.
2. The normal pulse width of each pulse is 0.05 µs to 1.0 µs.
3. For long ranges, a wide pulse and a slow repetition rate is used.
4. For short ranges, a narrow pulse and a fast repetition rate is used.

The last two statements are obvious when you think about them. If the radar is looking at something farther away, it will take longer for the signal to come back. Therefore, the repetition rate needs to be slower, to give time for the signal to return before the next pulse. Because more time is available, the individual pulses can also be longer.

From the facts above, you can answer most of the pulse repetition questions through the process of elimination. If the target is close, then the correct answer will be close to 0.05 µs for the pulse width, and close to 2000 for the repetition rate. If the target is far away, then the pulse width will be close to 1.0 µs, and the repetition rate will be close to 500 pulses per second.

Here are the answers to the questions on the exam. As you can see, you can get all of these answers by knowing the four facts shown above:

1.5 miles: 0.05 µs and 2000 pulses per second.

6 miles: 0.25 µs and 1000 pulses per second.

25 or 48 miles: 1.0 µs and 500 pulses per second.

There is one question about care of the magnetron. To protect the magnetron, one should keep metal tools away, not subject it to excessive heat, and not subject it to shocks and blows. For this question, you are asked to select the one item that has nothing to do with protecting the magnetron. That answer is "keep the TR properly tuned."

Since both the transmitter and receiver share the same antenna, the system will have a duplexer, which is an electronic switch allowing use of the shared antenna. The ATR box prevents the received signal from entering the transmitter.

At sea, an echo box can be used to determine the RADAR's performance.

A **waveguide** is typically used to transfer the RF energy from the RADAR to the antenna. Conductance through a waveguide takes place through electromagnetic and electrostatic fields in the walls of the waveguide. Energy is coupled into and out of the waveguide with a thin piece of wire as an antenna.

For the exam, you need to know some terminology relating to RADAR antennas. A **J-hook** is the pickup device is used to extract the RADAR signal from the wave guide A **slotted array** is a common type of shipboard RADAR antenna. As with any antenna which exhibits gain, the beamwidth of the antenna decreases as the gain of the antenna is increased.

The traveling wave tube of a RADAR is surrounded by a magnetic field. The purpose of this field is to prevent the electron beam from spreading.

A RADAR collision avoidance system on a ship is tied in with other instruments on the ship and receives data from them. The test asks you which instrument is **not** connected, and that is the anemometer. The other instruments tell the unit the ship's position and speed, and there is therefore no reason for the collision avoidance system to know the wind speed.

Finally, you need to know the abbreviation ARPA. This stands for "Automatic RADAR Plotting Aid."

3-9O01 What is the normal range of pulse repetition rates?

    A. 2,000 to 4,000 pps.

B.   1,000 to 3,000 pps.

C.   500 to 1,000 pps.

D.   500 to 2,000 pps.

3-9O02 The RADAR range in nautical miles to an object can be found by measuring the elapsed time during a RADAR pulse and dividing this quantity by:

A.   0.87 seconds.

B.   1.15 µs.

C.   12.346 µs.

D.   1.073 µs.

3-9O03 What is the normal range of pulse widths?

A.   .05 µs to 0.1 µs.

B.   .05 µs to 1.0 µs.

C.   1.0 µs to 3.5 µs.

D.   2.5 µs to 5.0 µs.

3-9O04 Shipboard RADAR is most commonly operated in what band?

A.   VHF.

B.   UHF.

C.   SHF.

D.   EHF.

3-9O05 The pulse repetition rate (prr) of a RADAR refers to the:

A. Reciprocal of the duty cycle.

B. Pulse rate of the local oscillator.

C. Pulse rate of the klystron.

D. Pulse rate of the magnetron.

3-9006 If the elapsed time for a RADAR echo is 62 microseconds, what is the distance in nautical miles to the object?

A. 5 nautical miles.

B. 87 nautical miles.

C. 37 nautical miles.

D. 11.5 nautical miles.

3-9101 The ATR box:

A. Prevents the received signal from entering the transmitter.

B. Protects the receiver from strong RADAR signals.

C. Turns off the receiver when the transmitter is on.

D. All of the above.

3-9102 What is the purpose or function of the RADAR duplexer/circulator? It is a/an:

A. Coupling device that is used in the transition from a rectangular waveguide to a circular waveguide.

B. Electronic switch that allows the use of one antenna for both transmission and reception.

C. Modified length of waveguide that is used to sample a portion of the transmitted energy for testing purposes.

D.   Dual section coupling device that allows the use of a magnetron as a transmitter.

3-9103 What device can be used to determine the performance of a RADAR system at sea?

A.   Echo box.

B.   Klystron.

C.   Circulator.

D.   Digital signal processor.

3-9104 What is the purpose of a synchro transmitter and receiver?

A.   Synchronizes the transmitted and received pulse trains.

B.   Prevents the receiver from operating during the period of the transmitted pulse.

C.   Transmits the angular position of the antenna to the indicator unit.

D.   Keeps the speed of the motor generator constant.

3-9106 The component or circuit providing the transmitter output power for a RADAR system is the:

A.   Thyratron.

B.   SCR.

C.   Klystron.

D.   Magnetron.

3-9201 When a RADAR is being operated on the 48 mile range setting, what is the most appropriate pulse width (PW) and pulse repetition rate (pps)?

A. 1.0 μs PW and 2,000 pps.

B. 0.05 μs PW and 2,000 pps.

C. 2.5 μs PW and 2,500 pps.

D. 1.0 μs PW and 500 pps.

3-9202 When a RADAR is being operated on the 6 mile range setting what is the most appropriate pulse width and pulse repetition rate?

A. 1.0 μs PW and 500 pps.

B. 2.0 μs PW and 3,000 pps.

C. 0.25 μs PW and 1,000 pps.

D. 0.01 μs PW and 500 pps.

3-9203 We are looking at a target 25 miles away. When a RADAR is being operated on the 25 mile range setting what is the most appropriate pulse width and pulse repetition rate?

A. 1.0 μs PW and 500 pps.

B. 0.25 μs PW and 1,000 pps.

C. 0.01 μs PW and 500 pps.

D. 0.05 μs PW and 2,000 pps.

3-9204 What pulse width and repetition rate should you use at long ranges?

A. Narrow pulse width and slow repetition rate.

B. Narrow pulse width and fast repetition rate.

C. Wide pulse width and fast repetition rate.

D. Wide pulse width and slow repetition rate.

3-9205 What pulse width and repetition rate should you use at short ranges?

    A. Wide pulse width and fast repetition rate.

    B. Narrow pulse width and slow repetition rate.

    C. Narrow pulse width and fast repetition rates.

    D. Wide pulse width and slow repetition rates.

3-9206 When a RADAR is being operated on the 1.5 mile range setting, what is the most appropriate pulse width and pulse repetition rate?

    A. 0.25 μs PW and 1,000 pps.

    B. 0.05 μs PW and 2,000 pps.

    C. 1.0 μs PW and 500 pps.

    D. 2.5 μs PW and 2,500 pps.

3-9302 What type of antenna or pickup device is used to extract the RADAR signal from the wave guide?

    A. J-hook.

    B. K-hook.

    C. Folded dipole.

    D. Circulator.

3-9303 What happens to the beamwidth of an antenna as the gain is increased? The beamwidth:

    A. Increases geometrically as the gain is increased.

B.   Increases arithmetically as the gain is increased.

C.   Is essentially unaffected by the gain of the antenna.

D.   Decreases as the gain is increased.

3-93O4 A common shipboard RADAR antenna is the:

A.   Slotted array.

B.   Dipole.

C.   Stacked Yagi.

D.   Vertical Marconi.

3-93O5 Conductance takes place in a waveguide:

A.   By interelectron delay.

B.   Through electrostatic field reluctance.

C.   In the same manner as a transmission line.

D.   Through electromagnetic and electrostatic fields in the walls of the waveguide.

3-93O6 To couple energy into and out of a waveguide use:

A.   Wide copper sheeting.

B.   A thin piece of wire as an antenna.

C.   An LC circuit.

D.   Capacitive coupling.

3-94O1 The permanent magnetic field that surrounds a traveling-wave tube (TWT) is intended to:

A. Provide a means of coupling.

B. Prevent the electron beam from spreading.

C. Prevent oscillations.

D. Prevent spurious oscillations.

3-9403 In the term "ARPA RADAR," ARPA is the acronym for which of the following?

A. Automatic RADAR Plotting Aid.

B. Automatic RADAR Positioning Angle.

C. American RADAR Programmers Association.

D. Authorized RADAR Programmer and Administrator.

3-9404 Which of the following is NOT a precaution that should be taken to ensure the magnetron is not weakened:

A. Keep metal tools away from the magnet.

B. Do not subject it to excessive heat.

C. Keep the TR properly tuned.

D. Do not subject it to shocks and blows.

3-9406 RADAR collision avoidance systems utilize inputs from each of the following except your ship's:

A. Gyrocompass.

B. Navigation position receiver.

C. Anemometer.

D. Speed indicator.

37: PULSE MODULATION

Some of the RADAR questions refer to pulse modulation, since RADAR is probably the most common use of pulse emissions. However, it can be used for other purposes, and there are a few questions relating to pulse modulation in general. In any form of pulse modulation, since the transmitter is emitting power for only a fraction of the duty cycle, the peak transmitter power will be much greater than the average power.

There are two methods of modulating a pulse wave that are covered on the test. The first of these is pulse width modulation, in which the modulating signal varies the width of the pulse. Pulse width modulation can even be used to transmit voice, by varying the duration of the pulse depending on the voice waveform at that point.

The other method is pulse position modulation, in which the modulating signal varies the time at which each pulse occurs.

Finally, you need to know that in pulse-code modulation using time division multiplex, it is important to synchronize the transmit and receive clock pulse rates.

3-59H1 What is an important factor in pulse-code modulation using time-division multiplex?

   A. **Synchronization of transmit and receive clock pulse rates.**

B. Frequency separation.

C. Overmodulation and undermodulation.

D. Slight variations in power supply voltage.

3-59H2 In a pulse-width modulation system, what parameter does the modulating signal vary?

A. Pulse frequency.

B. Pulse duration.

C. Pulse amplitude.

D. Pulse intensity.

3-59H3 What is the name of the type of modulation in which the modulating signal varies the duration of the transmitted pulse?

A. Amplitude modulation.

B. Frequency modulation.

C. Pulse-height modulation.

D. Pulse-width modulation.

3-59H4 Which of the following best describes a pulse modulation system?

A. The peak transmitter power is normally much greater than the average power.

B. Pulse modulation is sometimes used in SSB voice transmitters.

C. The average power is normally only slightly below the peak power.

D. The peak power is normally twice as high as the average power.

3-59H5 In a pulse-position modulation system, what parameter does the modulating signal vary?

    A.   The number of pulses per second.

    B.   **The time at which each pulse occurs.**

    C.   Both the frequency and amplitude of the pulses.

    D.   The duration of the pulses.

3-59H6 What is one way that voice is transmitted in a pulse-width modulation system?

    A.   A standard pulse is varied in amplitude by an amount depending on the voice waveform at that instant.

    B.   The position of a standard pulse is varied by an amount depending on the voice waveform at that instant.

    C.   **A standard pulse is varied in duration by an amount depending on the voice waveform at that instant.**

    D.   The number of standard pulses per second varies depending on the voice waveform at that instant.

38: SATELLITES

The test covers a number of satellite systems used for marine communication and navigation. The test has questions on Iridium, INMARSAT, COSPAS-SARSAT, and GPS.

The **Iridium** satellite system, originally developed by Motorola, provides communications through 66 low-earth orbit satellites at an altitude of 485 miles. The system provides digital voice and data communication at 2.4 kbps. Each satellite has 48 spot beams, each with a footprint of 30 miles in diameter.

COSPAS-SARSAT is an international satellite system monitoring for EPIRB (emergency position-indicating radio beacons) operating on 406 MHz. "SAR" in the name stands for "Search and Rescue". (COSPAS is a similar acronym in Russian.) COSPAS-SARSAT determines the location of the EPIRB by measuring the Doppler shift from the EPIRB.

INMARSAT is a British-based satellite telecommunications system with eleven geostationary satellites. In contrast to Iridium, its satellites are at a very high altitude, 22,177 miles. The satellites provide coverage in nearly all of the world's navigable waterways. A specific satellite is in place to provide coverage to specific areas, and these areas are designated with abbreviations such as AOR-E (Atlantic Ocean Region-East), AOR-W (Atlantic Ocean Region-West), POR (Pacific Ocean Region), and IOR (Indian Ocean Region).

When you remember these abbreviations, one of the exam questions is very easy. It gives a number of statements, and asks which one is correct. The correct answer is "AOR-E at 15.5° W, AOR-W at 54° W, POR at 178° E and IOR at 64.5° E." This statement is merely specifying the longitudes of the satellites covering the various ocean regions.

If you remember this, and you remember your geography, you can eliminate the wrong answers by the process of elimination. You should also remember that the Atlantic ocean is divided into two regions. This stands to reason because INMARSAT is British, and you would expect its best coverage in the Atlantic.

If you forget the longitudes, remember that Greenwich, England, is longitude zero, and each time zone is fifteen degrees of longitude. Therefore, the U.S. east coast, five time zones from Greenwich Mean Time, is approximately 75 degrees west longitude. The U.S. west coast, eight time zones from Greenwich Mean Time, is approximately 120 degrees west longitude.

From these facts, we can eliminate all of the wrong answers. In one wrong answer, AOR-W is at 85 degrees, which would put it over land, which would limit its use for maritime communication. The other two wrong answers have a single AOR, and instead have the Pacific or Indian oceans divided.

If you prefer to simply memorize the answer, remember that the correct answer is the only one that contains the number 15.5.

INMARSAT-B provides voice at 16 kbps, fax at 14.4 kbps, and high-speed data at 64/54. You should also know that INMARSAT-B voice communications, CODECs are used to digitize the voice signal.

INMARSAT-M has somewhat less bandwidth capability. It provides Voice at 6.2 kbps, data and fax at 2.4 kbps, and e-mail.

INMARSAT-C is a store-and-forward system, for both routine and distress communications. The INMARSAT mini-M system provides spot beams for small craft communication.

There is one question asking what conditions would make INMARSAT-B communications impossible. The correct answer is "all of the above", and most of these reasons should be obvious. For example, communication is not possible if the satellite is below the horizon, out of range, or behind an obstruction.

High speed data is available from three INMARSAT services. Those are B, M4, and Fleet.

An omnidirectional antenna is used for INMARSAT-C, and a larger directional antenna is used for INMARSAT-B. If you forget this, remember that INMARSAT-B provides high speed data and voice, whereas INMARSAT-C provides store-and-forward data. Obviously, the larger antenna is needed for the greater amount data.

Global Positioning Service (**GPS**) satellites orbit at 12,554 miles. Normally, 24 satellites are in operation. They are equally spaced in six orbital planes and inclined about 55 degrees. (If you wish to memorize these facts about GPS, just remember the numbers 6, 12, and 24: Six orbital planes, 12,000 miles, 24 satellites.) To get complete position and time data, the receiver must receive signals from four satellites. Differential GPS (DGPS) is a system that provides additional correction factors to improve position accuracy.

One GPS hardware question appears in two separate sections of the question pool, so it's possible you might get the nearly identical question twice. You should know that to verify the correct GPS sentence to a marine DSC VHF radio, you simply look for the latitude and longitude (and speed) on the display.

3-81L5 How might an installer verify correct GPS sentence to marine DSC VHF radio?

    A.   Press and hold the red distress button.

    B.   Look for latitude and longitude on the display.

    C.   Look for GPS confirmation readout.

    D.   Ask for VHF radio check position report.

3-85N6 How might an installer verify correct GPS sentence to marine DSC VHF radio?

A. Look for latitude and longitude, plus speed, on VHF display.

B. Press and hold the red distress button.

C. Look for GPS confirmation readout.

D. Ask for VHF radio check position report.

3-95P1 What is the orbiting altitude of the Iridium satellite communications system?

A. 22,184 miles.

B. 11,492 miles.

C. 4,686 miles.

D. 485 miles.

3-95P3 What services are provided by the Iridium system?

A. Analog voice and Data at 4.8 kbps.

B. Digital voice and Data at 9.6 kbps.

C. Digital voice and Data at 2.4 kbps.

D. Analog voice and Data at 9.6 kbps.

3-95P4 Which of the following statements about the Iridium system is true?

A. There are 48 spot beams per satellite with a footprint of 30 miles in diameter.

B. There are 48 satellites in orbit in 4 orbital planes.

C. The inclination of the orbital planes is 55 degrees.

D. The orbital period is approximately 85 minutes.

3-95P5 What is the main function of the COSPAS-SARSAT satellite system?

    A. Monitor 121.5 MHz for voice distress calls.

    B. Monitor 406 MHz for distress calls from EPIRBs.

    C. Monitor 1635 MHz for coded distress calls.

    D. Monitor 2197.5 kHz for hexadecimal coded DSC distress messages.

3-95P6 How does the COSPAS-SARSAT satellite system determine the position of a ship in distress?

    A. By measuring the Doppler shift of the 406 MHz signal taken at several different points in its orbit.

    B. The EPIRB always transmits its position which is relayed by the satellite to the Local User Terminal.

    C. It takes two different satellites to establish an accurate position.

    D. None of the above.

3-96P1 What is the orbital altitude of INMARSAT Satellites?

    A. 16, 436 miles.

    B. 22,177 miles.

    C. 10, 450 miles.

    D. 26,435 miles.

3-96P2 Which of the following describes the INMARSAT Satellite system?

    A. AOR at 35° W, POR-E at 165° W, POR-W at 155° E and IOR at 56.5° E.

B.   AOR-E at 25° W, AOR-W at 85° W, POR at 175° W and IOR at 56.5° E.

C.   AOR-E at 15.5° W, AOR-W at 54° W, POR at 178° E and IOR at 64.5° E.

D.   AOR at 40° W, POR at 178° W, IOR-E at 109° E and IOR-W at 46° E.

3-96P3   What are the directional characteristics of the INMARSAT-C SES antenna?

A.   Highly directional parabolic antenna requiring stabilization.

B.   Wide beam width in a cardioid pattern off the front of the antenna.

C.   Very narrow beam width straight-up from the top of the antenna.

D.   Omnidirectional.

3-96P4   When engaging in voice communications via an INMARSAT-B terminal, what techniques are used?

A.   CODECs are used to digitize the voice signal.

B.   Noise-blanking must be selected by the operator.

C.   The voice signal must be compressed to fit into the allowed bandwidth.

D.   The voice signal will be expanded at the receiving terminal.

3-96P5   Which of the following statements concerning INMARSAT geostationary satellites is true?

A.   They are in a polar orbit, in order to provide true global coverage.

B.   They are in an equatorial orbit, in order to provide true global coverage.

C.   They provide coverage to vessels in nearly all of the world's navigable waters.

D.   Vessels sailing in equatorial waters are able to use only one satellite, whereas other vessels are able to choose between at least two satellites.

3-96P6 Which of the following conditions can render INMARSAT -B communications impossible?

A.   An obstruction, such as a mast, causing disruption of the signal between the satellite and the SES antenna when the vessel is steering a certain course.

B.   A satellite whose signal is on a low elevation, below the horizon.

C.   Travel beyond the effective radius of the satellite.

D.   All of these.

3-97P1 What is the best description for the INMARSAT-C system?

A.   It provides slow speed telex and voice service.

B.   It is a store-and-forward system that provides routine and distress communications.

C.   It is a real-time telex system.

D.   It provides world-wide coverage.

3-97P2 The INMARSAT mini-M system is a:

A.   Marine SONAR system.

B.   Marine global satellite system.

C.   Marine depth finder.

D. Satellite system utilizing spot beams to provide for small craft communications.

3-97P3 What statement best describes the INMARSAT-B services?

A. Voice at 16 kbps, Fax at 14.4 kbps and high-speed Data at 64/54.

B. Store and forward high speed data at 36/48 kbps.

C. Voice at 3 kHz, Fax at 9.6 kbps and Data at 4.8 kbps.

D. Service is available only in areas served by highly directional spot beam antennas.

3-97P4 Which INMARSAT systems offer High Speed Data at 64/54 kbps?

A. C.

B. B and C.

C. Mini-M.

D. B, M4 and Fleet.

3-97P5 When INMARSAT-B and INMARSAT-C terminals are compared:

A. INMARSAT-C antennas are small and omni-directional, while INMARSAT-B antennas are larger and directional.

B. INMARSAT-B antennas are bulkier but omni-directional, while INMARSAT-C antennas are smaller and parabolic, for aiming at the satellite.

C. INMARSAT-B antennas are parabolic and smaller for higher gain, while INMARSAT-C antennas are larger but omni-directional.

D. INMARSAT-C antennas are smaller but omni-directional, while

INMARSAT-B antennas are parabolic for lower gain.

3-97P6 What services are provided by the INMARSAT-M service?

    A. Data and Fax at 4.8 kbps plus e-mail.

    B. Voice at 3 kHz, Fax at 9.6 kbps and Data at 4.8 kbps.

    C. Voice at 6.2 kbps, Data at 2.4 kbps, Fax at 2.4 kbps and e-mail.

    D. Data at 4.8 kbps and Fax at 9.6 kbps plus e-mail.

3-98P1 Global Positioning Service (GPS) satellite orbiting altitude is:

    A. 4,686 miles.

    B. 24,184 miles.

    C. 12,554 miles.

    D. 247 miles.

3-98P3 How many GPS satellites are normally in operation?

    A. 8

    B. 18

    C. 24

    D. 36

3-98P4 What best describes the GPS Satellites orbits?

    A. They are in six orbital planes equally spaced and inclined about 55 degrees to the equator.

    B. They are in four orbital planes spaced 90 degrees in a polar orbit.

C. They are in a geosynchronous orbit equally spaced around the equator.

D. They are in eight orbital planes at an altitude of approximately 1,000 miles.

3-98P5 How many satellites must be received to provide complete position and time?

    A. 1

    B. 2

    C. 3

    D. 4

3-98P6 What is DGPS?

    A. Digital Ground Position System.

    B. A system to provide additional correction factors to improve position accuracy.

    C. Correction signals transmitted by satellite.

    D. A system for providing altitude corrections for aircraft.

39: REPAIR AND MAINTENANCE, TEST
EQUIPMENT

The exam has a few questions on repair techniques, and quite a few
questions about various kinds of test equipment.

It asks for the best tool to use for stripping wire. My personal favorite is
the pocket knife, but according to the FCC, that is the wrong answer.
Instead, the ideal method of stripping wire, according to the FCC, is the
thermal stripper.

To remove solder from a printed circuit board, the best tool is what the
FCC calls a "vacuum device", which is commonly called a "solder
sucker".

Plastic wire ties should be cut flush using flush-cut diagonal pliers. And
speaking of tie wraps, white or translucent tie wraps should not be used
on a radio tower because UV radiation from the sun deteriorates the
plastic.

A hot gas bonder is used for non-contact melting of solder.

If you're ever installing a rack-mounted piece of equipment, you want to
mount it level. When using a carpenter's level, you want to be just inside
the bubble.

One troubleshooting question asks what might cause low voltage on a marine SSB transmitter while transmitting. The correct answer is that the negative fuse (black wire) is blown. The giveaway for this question is that the problem shows up while transmitting, when the radio is drawing more current. This hints that the problem is due to a high resistance somewhere, which is causing a large voltage drop when the current is high. If the negative fuse blows, the radio is probably grounded elsewhere, such as through the antenna ground. Therefore, the problem might not be readily apparent, since the receiver continues to work normally. But since the accidental ground connection probably has a higher resistance than the normal DC ground, the voltage drop becomes more apparent when transmitting.

An oscilloscope is a test instrument used to view voltage waveforms. The vertical axis is typically voltage, and the horizontal axis is typically time. One question asks you to name something you cannot do with an oscilloscope. The answer is that you can't use an oscilloscope to measure the speed of light.

The maximum frequency response of an oscilloscope is determined by the vertical amplifier. The accuracy, frequency response, and stability are determined by the sweep oscillator and the deflection amplifier bandwidth.

To decrease circuit loading when hooking up an oscilloscope, you could use a divider probe. In the question on the exam, this is a 10:1 divider probe.

A spectrum analyzer has a screen like an oscilloscope, but has a different function. With a spectrum analyzer, you are looking at a piece of the RF spectrum, and the instrument shows you how strong the RF signals are on different frequencies within that range. In other words, the horizontal axis shows frequency, and the vertical axis shows amplitude. A spectrum analyzer is useful for displaying spurious signals in a transmitter's output.

As you probably know, a frequency counter measures frequency, and that is the correct answer to one question. In putting together the test, the FCC apparently realized that this question was too easy, and additional verbiage was added to make the question appear more complicated than it really is. To do this, they added the definition of frequency, "the time between events, or the frequency, which is the reciprocal of the time." This is still the right answer. A prescaler can be used at the input of a frequency counter. The prescaler divides the frequency of a high frequency signal so that it can be measured on a low frequency counter.

A frequency standard is merely a device used to produce a highly accurate reference frequency. A rubidium standard as the time standard is often used.

3-40E1 What is the purpose of a prescaler circuit?

    A. Converts the output of a JK flip-flop to that of an RS flip-flop.

    B. Multiplies an HF signal so a low-frequency counter can display the operating frequency.

    C. Prevents oscillation in a low frequency counter circuit.

    D. Divides an HF signal so that a low-frequency counter can display the operating frequency.

3-75L1 How is a frequency counter used?

    A. To provide reference points on an analog receiver dial thereby aiding in the alignment of the receiver.

    B. To heterodyne the frequency being measured with a known variable frequency oscillator until zero beat is achieved, thereby indicating the unknown frequency.

    C. To measure the deviation in an FM transmitter in order to determine the percentage of modulation.

    D. To measure the time between events, or the frequency, which is the reciprocal of the time.

3-75L2 What is a frequency standard?

    A. A well-known (standard) frequency used for transmitting certain messages.

    B. A device used to produce a highly accurate reference frequency.

C. A device for accurately measuring frequency to within 1 Hz.

D. A device used to generate wide-band random frequencies.

3-75L5 Which of the following frequency standards is used as a time base standard by field technicians?

A. Quartz Crystal.

B. Rubidium Standard.

C. Cesium Beam Standard.

D. LC Tank Oscillator.

3-76L1 What is used to decrease circuit loading when using an oscilloscope?

A. Dual input amplifiers.

B. 10:1 divider probe.

C. Inductive probe.

D. Resistive probe.

3-76L2 How does a spectrum analyzer differ from a conventional oscilloscope?

A. The oscilloscope is used to display electrical signals while the spectrum analyzer is used to measure ionospheric reflection.

B. The oscilloscope is used to display electrical signals in the frequency domain while the spectrum analyzer is used to display electrical signals in the time domain.

C. The oscilloscope is used to display electrical signals in the time domain while the spectrum analyzer is used to display electrical signals in the frequency domain.

D. The oscilloscope is used for displaying audio frequencies and the spectrum analyzer is used for displaying radio frequencies.

3-76L3 What stage determines the maximum frequency response of an oscilloscope?

A. Time base.

B. Horizontal sweep.

C. Power supply.

D. Vertical amplifier.

3-76L4 What factors limit the accuracy, frequency response, and stability of an oscilloscope?

A. **Sweep oscillator quality and deflection amplifier bandwidth.**

B. Tube face voltage increments and deflection amplifier voltage.

C. Sweep oscillator quality and tube face voltage increments.

D. Deflection amplifier output impedance and tube face frequency increments.

3-76L5 An oscilloscope can be used to accomplish all of the following except:

A. Measure electron flow with the aid of a resistor.

B. Measure phase difference between two signals.

C. **Measure velocity of light with the aid of a light emitting diode.**

D. Measure electrical voltage.

3-77L2 What does the horizontal axis of a spectrum analyzer display?

A.   Amplitude.

B.   Voltage.

C.   Resonance.

D.   Frequency.

3-77L3 What does the vertical axis of a spectrum analyzer display?

A.   Amplitude.

B.   Duration.

C.   Frequency.

D.   Time.

3-77L5 What test instrument can be used to display spurious signals in the output of a radio transmitter?

A.   A spectrum analyzer.

B.   A wattmeter.

C.   A logic analyzer.

D.   A time domain reflectometer.

3-80L3 Why should you not use white or translucent plastic tie wraps on a radio tower?

A.   White tie wraps are not FAA approved.

B.   UV radiation from the Sun deteriorates the plastic very quickly.

C.   The white color attracts wasps

D.   The black tie wraps may cause electrolysis.

3-79L2 Which of the following is the preferred method of cleaning solder from plated-through circuit-board holes?

A. Use a dental pick.

B. Use a vacuum device.

C. Use a soldering iron tip that has a temperature above 900 degrees F.

D. Use an air jet device.

3-79L3 What is the proper way to cut plastic wire ties?

A. With scissors.

B. With a knife.

C. With semi-flush diagonal pliers.

D. With flush-cut diagonal pliers and cut flush.

3-79L4 The ideal method of removing insulation from wire is:

A. The thermal stripper.

B. The pocket knife.

C. A mechanical wire stripper.

D. The scissor action stripping tool.

3-79L5 A "hot gas bonder" is used:

A. To apply solder to the iron tip while it is heating the component.

B. For non-contact melting of solder.

C. To allow soldering both sides of the PC board simultaneously.

D.   To cure LCA adhesives.

3-80L5 What tolerance off of plumb should a single base station radio rack be installed?

A.   No tolerance allowed.

B.   Just outside the bubble on a level.

C.   All the way to one end.

D.   Just inside the bubble on a level.

3-86N2 What might contribute to apparent low voltage on marine SSB transmitting?

A.   Blown red fuse.

B.   Too much grounding.

C.   Blown black negative fuse.

D.   Antenna mismatch.

40: INDUSTRY STANDARDS AND SPECIFICATIONS

NMEA stands for the National Marine Electronics Association, and the test covers two NMEA electrical and data specifications. NMEA 0183 is an older standard, and is being phased out in favor of NMEA 2000.

There is a single question on the test about NMEA 0183. NMEA 0183 is mentioned in the question, which asks how a data line should be grounded. The answer is that it should be terminated to ground at the talker and unterminated at the listener. This is the only question mentioning NMEA 0183. Remember that NMEA is **never** the correct **answer** to any question. It is mentioned only in the question.

The newer standard is NMEA 2000. One question asks "what data language is bi-directional, multi-transmitter, multi-receiver network?" The answer is NMEA 2000. You also need to know the following facts regarding NMEA 2000 specifications:

1. NMEA 2000 is designed so that if one device in line should fail, there will be no interruption to other devices.

2. A "load equivalence number" (LEN) is equivalent to 50 mA. (If you forget this number, think of a device such as a VHF transceiver, whose current draw while squelched is probably about 50 mA.)

3. A system with devices at a single location may be powered

through an end-powered network.

4.   A voltage drop of 1.5 volts at the end of the last segment will satisfy cabling plans. (Think of an automotive alternator putting out 13.7 volts. Everything will work fine if 12.2 volts is available at the end of the line.)

P25 is a protocol for trunking of land mobile radio systems. There is one question on the test regarding P25, and you would probably get it right by making an educated guess. The question asks whether a P25 radio system can be monitored with a scanner. The correct answer is that it can, if the scanner has P25 decoding.

The test also asks some questions regarding color coding and cable types. If you're not familiar with these schemes, you will need to memorize the following:

1. In a 100-pair cable, the binder for pairs 51-75 is green.

2. In a 25-pair switchboard cable, the sixth pair is red/blue, blue/red. (This is the only answer with only two colors).

3. To connect an SSB automatic tuner to an insulated backstay, you would use GTO-15 high-voltage cable. (This identical question appears twice in the question pool, so you might actually see it twice on the exam.)

The most common code for exchanging of data from one computer to another is ASCII.

Finally, you need to know two things about the International Organization for Standardization's model for packet radio communications. You need to know that the physical layer is responsible for the actual transmission of data. And you need to know that the link layer arranges the bits into frames and controls the data flow. As the name implies, the physical layer is the actual physical transmission through a physical radio. So if you get the question mentioning actual transmission, then the answer is physical. If you get the other question, then the answer is link.

3-78L1 Can a P25 radio system be monitored with a scanner?

A. Yes , regardless if it has P25 decoding or not.

B. No.

C. **Yes, if the scanner has P25 decoding.**

D. Yes, but it must also have P26 decoding.

3-80L1 What color is the binder for pairs 51-75 in a 100-pair cable?

A. Red

B. Blue

C. Black

D. **Green**

3-80L4 What is the 6th pair color code in a 25 pair switchboard cable as is found in building telecommunications interconnections?

A. Blue/Green, Green/Blue.

B. Red/Blue, White/Violet.

C. **Red/Blue, Blue/Red.**

D. White/Slate, Slate/White.

3-80L6 What type of wire would connect an SSB automatic tuner to an insulated backstay?

A. **GTO-15 high-voltage cable.**

B. RG8U.

C. RG213.

D. 16-gauge two-conductor.

3-83M3 Which of the following codes has gained the widest acceptance for exchange of data from one computer to another?

A. Gray.

B. Baudot.

C. Morse.

D. ASCII.

3-83M4 The International Organization for Standardization has developed a seven-level reference model for a packet-radio communications structure. What level is responsible for the actual transmission of data and handshaking signals?

A. The physical layer.

B. The transport layer.

C. The communications layer.

D. The synchronization layer.

3-83M5 What CODEC is used in Phase 2 P25 radios?

A. IWCE

B. IMBC

C. IMMM

D. AMBE

3-83M6 The International Organization for Standardization has developed a seven-level reference model for a packet-radio communications structure. The _____ level arranges the bits into frames and controls data flow.

A. Transport layer.

B. Link layer.

C. Communications layer.

D. Synchronization layer.

3-86N3 What type of wire connects an SSB automatic tuner to an insulated backstay?

A. RG8U.

B. RG213.

C. 16-gauge two-conductor.

D. GTO-15 high-voltage cable.

3-89N1 What data language is bi-directional, multi-transmitter, multi-receiver network?

A. NMEA 2000.

B. NMEA 0181.

C. NMEA 0182.

D. NMEA 0183.

3-89N2 How should shielding be grounded on an NMEA 0183 data line?

A. Unterminated at both ends.

B. Terminated to ground at the talker and unterminated at the listener.

C. Unterminated at the talker and terminated at the listener.

D. Terminated at both the talker and listener.

3-89N3 What might occur in NMEA 2000 network topology if one device in line should fail?

    A.   The system shuts down until the device is removed.

    B.   Other electronics after the failed device will be inoperable.

    C.   The main fuse on the backbone may open.

    D.   There will be no interruption to all other devices.

3-89N4 In an NMEA 2000 device, a load equivalence number (LEN) of 1 is equivalent to how much current consumption?

    A.   50 mA

    B.   10 mA

    C.   25 mA

    D.   5 mA

3-89N5 An NMEA 2000 system with devices in a single location may be powered using this method:

    A.   Dual mid-powered network.

    B.   End-powered network.

    C.   Individual devices individually powered.

    D.   No 12 volts needed for NMEA 2000 devices.

3-89N6 What voltage drop at the end of the last segment will satisfy NMEA 2000 network cabling plans?

    A.   0.5 volts

    B.   2.0 volts

C. 1.5 volts

D. 3.0 volts

41: MISCELLANEOUS

There are a few questions on the test that don't fit well into the previous chapters. In some cases, there is only a single question on a given topic, and there's really no need to have any in-depth knowledge.

PCS (Personal Communication System) refers to portable electronic devices such as cellular phones operating in the 1900 MHz band. All you need to know for the test about PCS is that the coding systems **CDMA** and **GSM** are used. You also need to know that the RF power reading on a CDMA radio is not very accurate if using an analog power meter.

TDMA stands for time division multiple access. As long as you remember what the abbreviation stands for, you will have no problem with the single question on this topic. In fact, the question gives away the answer by asking how TDMA can carry multiple conversations sequentially. As soon as you spot the word sequentially, the correct answer is obvious: TDMA uses separate time slots for each user.

CODEC stands for "coder/decoder". When asked what it is, look for the only answer that includes the word "coder/decoder": a coder/decoder IC or circuit that converts a voice signal to digital format. Also, you need to know that GSM uses an RPE type CODEC.

Just as with your TV remote control at home, infra-red communications is often used to program radios.

A **SINAD** meter is the most common method of determining the exact sensitivity of a receiver. You do this by measuring the recovered audio for 12 dB of SINAD. As with all things relating to dB's, the number 3 is a magic number (representing a doubling), and many important dB ratios, such as this one, are exact multiples of 3. The correct answer, 12 dB, is the only one that is a multiple of 3. A SINAD meter consists of an AF voltmeter calibrated in dB, and a sharp 1000 Hz filter.

3-75L6 Which of the following contains a multirange AF voltmeter calibrated in dB and a sharp, internal 1000 Hz bandstop filter, both used in conjunction with each other to perform quieting tests?

   A. SINAD meter.

   B. Reflectometer.

   C. Dip meter.

   D. Vector-impedance meter.

3-78L2 Which of the following answers is true?

   A. The RF Power reading on a CDMA (code division multiple access) radio will be very accurate on an analog power meter.

   B. The RF Power reading on a CDMA radio is not accurate on an analog power meter.

   C. Power cannot be measured using CDMA modulation.

   D. None of the above.

3-78L3 What is a common method used to program radios without using a "wired" connection?

   A. Banding.

   B. Using the ultraviolet from a programmed radio to repeat the programming in another.

C. Infra-red communication.

D. Having the radio maker send down a programming signal via satellite.

3-78L4 What is the common method for determining the exact sensitivity specification of a receiver?

A. Measure the recovered audio for 12 dB of SINAD.

B. Measure the recovered audio for 10 dB of quieting.

C. Measure the recovered audio for 10 dB of SINAD.

D. Measure the recovered audio for 25 dB of quieting.

3-82M4 A TDMA radio uses what to carry the multiple conversations sequentially?

A. Separate frequencies.

B. Separate pilot tones.

C. Separate power levels.

D. Separate time slots.

3-82M6 What are the two most-used PCS (Personal Communications Systems) coding techniques used to separate different calls?

A. QPSK and QAM.

B. CDMA and GSM.

C. ABCD and SYZ.

D. AM and Frequency Hopping.

3-83M1 What is a CODEC?

    A. A device to read Morse code.

    B. A computer operated digital encoding compandor.

    C. A coder/decoder IC or circuitry that converts a voice signal into a predetermined digital format for encrypted transmission.

    D. A voice amplitude compression chip.

3-83M2 The GSM (Global System for Mobile Communications) uses what type of CODEC for digital mobile radio system communications?

    A. Regular-Pulse Excited (RPE).

    B. Code-Excited Linear Predictive (CLEP).

    C. Multi-Pulse Excited (MPE).

    D. Linear Excited Code (LEC).

3-84M6 What principle allows multiple conversations to be able to share one radio channel on a GSM channel?

    A. Frequency Division Multiplex.

    B. Double sideband.

    C. Time Division Multiplex.

    D. None of the above.

ABOUT THE AUTHOR

Richard Clem, WØIS, holds the GROL and Extra class amateur radio license. His previous works include study guides for other commercial and amateur license exams, and a multi-band antenna construction article published in *QST*. He is married to the illustrator, Yippy Clem, KCØOIA, and resides in St. Paul, Minnesota. He is an attorney in private practice.

www.ingramcontent.com/pod-product-compliance
Lightning Source LLC
Chambersburg PA
CBHW071358170526
45165CB00001B/98